PURE AND APPLIED MATHEMATICS

A Program of Monographs, Textbooks, and Lecture Notes

LECTURE NOTES
IN PURE AND APPLIED MATHEMATICS

1. *N. Jacobson,* Exceptional Lie Algebras
2. *L.-Å. Lindahl* and *F. Poulsen,* Thin Sets in Harmonic Analysis
3. *I. Satake,* Classification Theory of Semi-Simple Algebraic Groups
4. *F. Hirzebruch, W. D. Neumann,* and *S. S. Koh,* Differentiable Manifolds and Quadratic Forms
5. *I. Chavel,* Riemannian Symmetric Spaces of Rank One
6. *R. B. Burckel,* Characterization of C(X) among Its Subalgebras
7. *B. R. McDonald, A. R. Magid,* and *K. C. Smith,* Ring Theory: Proceedings of the Oklahoma Conference
8. *Yum-Tong Siu,* Techniques of Extension of Analytic Objects
9. *S. R. Caradus, W. E. Pfaffenberger, and Bertram Yood,* Calkin Algebras and Algebras of Operators on Banach Spaces
10. *Emilio O. Roxin, Pan-Tai Liu,* and *Robert L. Sternberg,* Differential Games and Control Theory
11. *Morris Orzech* and *Charles Small,* The Brauer Group of Commutative Rings
12. *S. Thomeier,* Topology and its Applications
13. *Jorge M. López* and *Kenneth A. Ross,* Sidon Sets

Other volumes in preparation

SIDON SETS

MARCEL DEKKER, INC.

270 Madison Avenue, New York, New York 10016

LIBRARY OF CONGRESS CATALOG CARD NUMBER: 75-193

ISBN: 0-8247-6289-4

Current printing (last digit):
10 9 8 7 6 5 4 3 2 1

PRINTED IN THE UNITED STATES OF AMERICA

PREFACE

The study of lacunary sets has played a significant and fascinating role in harmonic analysis. The classical setting involves sets of integers, but more recent studies have been geared at various levels of generality. The setting of discrete abelian groups is attractive because few extra difficulties are encountered in this generality and because the essential features of the theory are more apparent. The concept of Sidon set was isolated and given a life of its own nearly twenty years ago, but the most surprising results have been proved in the last five years. The principal innovators have been S. W. Drury and M. Déchamps-Gondim. It seems time to organize an account of Sidon sets that takes these recent developments into account. This is the aim of these notes.

Many of the authors whose papers are cited have helped us in one way or another. Several kept us posted on developments that are discussed, often too briefly, in Chapter 10. To all we offer our thanks. We also thank Mr. Chung Lin who read the complete manuscript and made useful suggestions. Finally, we thank our families for tolerating us during this project.

CONTENTS

SIDON SETS

Chapter 1

INTRODUCTION

Classical lacunary sets of integers (herein termed Hadamard sets) have been studied for over fifty years from the point of view of Fourier analysis. In a series of papers published from 1926 to 1941, Sidon showed that the lacunary sets possessed various interesting properties. In the 1950s and early 1960s, some of these properties were studied more closely and finally became definitions. Since these properties were more functional analytic in nature, while the original Hadamard property is arithmetic in nature, it was no longer necessary to restrict attention to sets of integers. Lacunary sets have since been studied in the duals of compact groups and in even more general settings. Apparently Kahane [1957] first used the term "Sidon set" and the modern point of view was well established with the appearance of Rudin's book [R] in 1962. Pioneer work was done by Stechkin [1956], Hewitt and Zuckerman [1959] and Rudin [1960]. For more historical comments, see the Notes to Section 37 in [HR; Vol. II] and Chapter 10.

Our understanding of Sidon sets was jolted in 1970 with the appearance of Drury [1970] and Déchamps-Gondim [1970a], who showed that Sidon sets are better behaved than had previously been suspected. Edwards, Hewitt and Ross [1972c] studied more special sets, so-called Fatou-Zygmund sets, only to have Drury [1974] show that all Sidon sets are Fatou-Zygmund sets. This means that Sidon sets enjoy the various properties established by Edwards, Hewitt and Ross and also that many older results can be easily improved. Moreover, the theory can now be efficiently organized to include these improvements. The purpose of these notes is to give such an account in the setting of compact (and discrete) abelian groups.

The main part of the notes is a reorganized account of the results in the four papers cited in the preceding paragraph. Of course, many other important contributions are also included. About forty-five percent of the material consists of results that are inessential to the main development. These results will be tagged by an asterisk and may be omitted. Since all the examples fall into this category, however, we suggest that readers at least glance through these items.

The main results mentioned above give affirmative answers to some of the questions posed on page 67 of [LP]. Problem 5.4 [LP] is answered by the Hartman-Wells Corollary 4.8. Problem 5.5 [LP] is answered by Drury's Theorem 3.3. And the first part of Problem 5.6 [LP] is answered by Déchamps-Gondim's Corollary 8.19. In fact, more is true; see her Corollary 8.23. It should be emphasized that these results are new in the classical setting of the circle group $\mathbb{T}$.

These notes assume familiarity with the basic parts of harmonic analysis over compact abelian groups as presented, for example, in [HR; mainly Chap. 8, Vol. II] or in [R; mainly Chaps. 1 and 2]. No previous knowledge of lacunarity is presumed, however.

1.1 NOTATION AND TERMINOLOGY. Throughout these notes G will denote a compact abelian group and X will denote its character group. Trig(G) denotes the space of complex-valued trigonometric polynomials on G. $C(G)$ denotes the space of complex-valued continuous functions on G. $L^p(G)$ denotes the usual Lebesgue space of p-th power integrable complex-valued functions on G constructed relative to normalized Haar measure λ on G. $M(G)$ denotes the space of complex regular Borel measures on G. For a subset S of G, $M(S)$ denotes the set of all $\mu \in M(G)$ such that $\mathrm{Supp}(|\mu|) \subset S$, where Supp signifies "support". For any set E, $\ell^p(E)$ for $p \in [1, \infty]$ and $c_0(E)$ have their usual meaning. For any number p in $[1, \infty]$, p' denotes the conjugate exponent $p/(p-1)$ (= ∞ if $p = 1$). $A(G)$ denotes the space of all (necessarily continuous) functions f on G whose Fourier transforms $\hat{f}$ belong to $\ell^1(X)$; the norm in this space is given by $\|f\|_A = \|f\|_{A(G)} = \|\hat{f}\|_1$.

Consider a subset E of X. A function f [or measure μ] on G is called E-<u>spectral</u> provided its Fourier transform $\hat{f}$ [or Fourier-Stieltjes transform $\hat{\mu}$] vanishes off E. If ϕ is a function on X, then $\phi|_E$ denotes its restriction to E. A subscript E on a space of functions or measures restricts that space to its E-spectral members. Thus $\mathrm{Trig}_E(G)$ consists of all the E-spectral trigonometric polynomials on G and $M_E(G)$ consists of all E-spectral measures on G. Similarly, subscripts and superscripts r and $+$ restrict the spaces to their real-valued and nonnegative real-valued members, respectively.

The sets of real numbers, the circle group, and the integers will be denoted by $\mathbb{R}$, $\mathbb{T}$, and $\mathbb{Z}$, respectively. Moreover, $\mathbb{Z}^+$ and $\mathbb{R}^+$ denote the sets of positive integers and positive real numbers, respectively.

1.2 DEFINITION. A subset E of X is called a <u>Sidon set</u> if there exists a constant $\kappa > 0$ so that to each $\phi \in \ell^{\infty}(E)$ there corresponds $\mu \in M(G)$ such that $\phi = \hat{\mu}|_E$ and $\|\mu\| \leq \kappa \|\phi\|_{\infty}$. Any constant κ with this property will be called a <u>Sidon constant</u> for E.

We next show that several properties are equivalent to Sidonicity. Any of them could serve as the definition. We have selected the one that seems to be used the most often. The following theorem is the first result established in all the standard treatments of Sidon sets; see [DR; p. 47], [E; 15.1.4], [HR; 37.2], [K; p. 127], [LP; p. 51], [R; 5.7.3]. For the sake of completeness, we also give a proof.

1.3 THEOREM. <u>For a subset</u> E <u>of</u> X, <u>the following are equivalent</u>:

(i) E <u>is a Sidon set with Sidon constant</u> κ;

(ii) <u>each</u> $\phi \in \ell^{\infty}(E)$ <u>has the form</u> $\hat{\mu}|_E$ <u>for some</u> μ <u>in</u> $M(G)$;

(iii) <u>given</u> $\phi : E \to \{-1,1\}$, <u>there is some</u> $\mu \in M(G)$ <u>such that</u> $\sup_{\chi \in E} |\hat{\mu}(\chi) - \phi(\chi)| < 1$;

(iv) <u>each</u> $\phi \in c_0(E)$ <u>has the form</u> $\hat{f}|_E$ <u>for some</u> f <u>in</u> $L^1(G)$;

(v) $L^{\infty}_E(G) \subset A(G)$;

(vi) $C_E(G) \subset A(G)$;

(vii) <u>there is a constant</u> $\varkappa > 0$ <u>so that</u> $\|f\|_A \leq \varkappa\|f\|_u$ <u>for all</u> $f \in \mathrm{Trig}_E(G)$.

<u>Proof</u>. We will prove (i) ⇒ (ii) ⇒ (iii) ⇒ (v) ⇒ (vi) ⇒ (vii) ⇒ (i) and (ii) ⇒ (iv) ⇒ (vii).

<u>(i) ⇒ (ii) ⇒ (iii)</u> are obvious.

<u>(iii) ⇒ (v)</u>. For any function $f \in L^1(G)$, f^* denotes the usual adjoint of f satisfying $(f^*)^\wedge = \overline{\hat{f}}$. Given $f \in L_E^\infty(G)$, we can write $f = f_1 + if_2$ where $f_1 = \frac{1}{2}(f + f^*)$ and $if_2 = \frac{1}{2}(f - f^*)$. It is easy to see that $\hat{f}_1$ and $\hat{f}_2$ are real valued and that $f_1, f_2 \in L_E^\infty(G)$. Since it suffices to show that f_1 and f_2 are in $A(G)$, we lose no generality in supposing that $\hat{f}$ is real valued. Now define $\phi : E \to \{-1,1\}$ so that $\phi\hat{f} = |\hat{f}|$. By (iii), there is a measure $\mu \in M(G)$ satisfying $\sup_{\chi\in E}|\hat{\mu}(\chi) - \phi(\chi)| < 1$. Hence for some $\delta > 0$, we have

$$|\hat{\mu}(\chi) - \phi(\chi)| \leq 1 - \delta \quad \text{for} \quad \chi \in E. \tag{1}$$

If $\nu = \frac{1}{2}(\mu + \mu^*)$, then $\hat{\nu}$ is the real part of $\hat{\mu}$ and $\hat{\nu}$ also satisfies (1). The function $g = \nu * f$ belongs to $L_E^\infty(G)$ and

$$\begin{aligned} |\hat{g}(\chi) - |\hat{f}(\chi)|| &= |\hat{\nu}(\chi)\hat{f}(\chi) - \phi(\chi)\hat{f}(\chi)| \\ &= |\hat{\nu}(\chi) - \phi(\chi)|\cdot|\hat{f}(\chi)| \leq (1 - \delta)|\hat{f}(\chi)| \end{aligned}$$

for $\chi \in E$. It follows that

$$\hat{g}(\chi) \geq \delta|\hat{f}(\chi)| \quad \text{for} \quad \chi \in E.$$

Let $\{h_\alpha\}$ be an approximate unit for $L^1(G)$ where each h_α belongs to $\mathrm{Trig}^+(G)$, $\hat{h}_\alpha \geq 0$ and $\|h_\alpha\|_1 = 1$ for all α; see [HR; 28.53]. For finite sets $F \subset E$, we have

$$\begin{aligned} \delta \sum_{\chi\in F} |\hat{f}(\chi)|\hat{h}_\alpha(\chi) &\leq \sum_{\chi\in F} \hat{g}(\chi)\hat{h}_\alpha(\chi) \leq g*h_\alpha(0) \\ &\leq \|g*h_\alpha\|_\infty \leq \|g\|_\infty\|h_\alpha\|_1 = \|g\|_\infty. \end{aligned}$$

Since $\lim_\alpha \hat{h}_\alpha(\chi) = 1$ for each $\chi \in X$, we conclude that

$$\delta \sum_{\chi \in F} |\hat{f}(\chi)| \leq \|g\|_\infty .$$

Since F is an arbitrary finite set in E, we conclude that $\hat{f} \in \ell^1(X)$ and $f \in A(G)$.

(v) ⇒ (vi) is obvious and (vi) ⇒ (vii) follows from the open mapping theorem applied to the inverse of the map $f \to \hat{f}|_E$ of $C_E(G)$ onto $\ell^1(E)$.

(vii) ⇒ (i). Consider fixed ϕ in $\ell^\infty(E)$. By (vii), we have $\|\hat{f}\|_1 \leq \kappa \|f\|_u$ for $f \in \mathrm{Trig}_E(G)$. Hence

$$L(f) = \sum_{\chi \in E} \hat{f}(\chi)\overline{\phi(\chi)}$$

defines a linear functional L on the subspace $\mathrm{Trig}_E(G)$ of $C(G)$ that is bounded: $|L(f)| \leq \kappa \|\phi\|_\infty \|f\|_u$ for $f \in \mathrm{Trig}_E(G)$. By the Hahn-Banach theorem and Riesz representation theorem, there is a measure $\mu \in M(G)$ such that $\|\mu\| \leq \kappa \|\phi\|_\infty$ and $L(f) = \overline{\int_G \bar{f}\, d\mu}$ for $f \in \mathrm{Trig}_E(G)$. Putting $f = \chi \in E$, we obtain $\hat{\mu}(\chi) = \phi(\chi)$.

(ii) ⇒ (iv). An application of the open mapping theorem ([HR; E.2]) to the map $\mu \to \hat{\mu}|_E$ of $M(G)$ onto $\ell^\infty(E)$ shows that there is a constant $\kappa > 0$ so that to each $\phi \in \ell^\infty(E)$ there corresponds a measure $\mu \in M(G)$ satisfying $\hat{\mu}|_E = \phi$ and $\|\mu\| \leq \kappa \|\phi\|_\infty$. Now let $\phi \in c_o(E)$ and assume that $\|\phi\|_\infty = 1$. For each $n \in \mathbb{Z}^+$, consider the finite set

$$E_n = \{\chi \in E : 2^{-n} < |\phi(\chi)| \leq 2^{-n+1}\}$$

and let ϕ_n be defined to be equal to ϕ on E_n and to be zero elsewhere on E. Now choose measures $\mu_n \in M(G)$ so that $\hat{\mu}_n|_E = \phi_n$ and $\|\mu_n\| \leq \kappa 2^{-n+1}$ for each $n \in \mathbb{Z}^+$. Since E_n is a finite set, we may choose $f_n \in \mathrm{Trig}(G)$ so that $\hat{f}_n(\chi) = 1$

for $\chi \in E_n$ and $\|f_n\|_1 \leq 2$ for each $n \in \mathbb{Z}^+$; see [HR; 31.37] or [R; 2.6.8]. Since $\Sigma_{n=1}^{\infty} \|f_n * \mu_n\|_1 \leq \varkappa \Sigma_{n=1}^{\infty} 2^{-n+2} < \infty$, the series $\Sigma_{n=1}^{\infty} f_n * \mu_n$ converges to a function $f \in L^1(G)$ and it is clear that $\hat{f}|_E = \phi$.

An alternate proof of <u>(ii) ⇒ (iv)</u> uses the Cohen factorization theorem [HR; 32.22] applied to the $L^1(G)$-module $c_o(E)$ where the module operation $c_o(E) \times L^1(G) \to c_o(E)$ is $(\psi, g) \to \psi \cdot \hat{g}|_E$. In fact, if ϕ is in $c_o(E)$, then ϕ has the form $\psi \cdot \hat{g}|_E$ where $\psi \in c_o(E)$ and $g \in L^1(G)$. By (ii), $\psi = \hat{\mu}|_E$ for some $\mu \in M(G)$. Then $\mu * g$ is in $L^1(G)$ and

$$(\mu * g)^\wedge|_E = \psi \cdot \hat{g}|_E = \phi.$$

<u>(iv) ⇒ (vii)</u>. This time we apply the open mapping theorem to the map $g \to \hat{g}|_E$ of $L^1(G)$ onto $c_o(E)$. It follows that there is a constant $\varkappa$ such that to each ϕ in $c_o(E)$ there corresponds a function g in $L^1(G)$ satisfying

$$\hat{g}|_E = \phi \quad \text{and} \quad \|g\|_1 \leq \varkappa \|\phi\|_\infty.$$

Now consider $f \in \mathrm{Trig}_E(G)$ and let $\phi \in c_o(E)$ be chosen so that $\phi(\chi)\hat{f}(\chi) = |\hat{f}(\chi)|$ for $\chi \in E$. Clearly we have $\|\phi\|_\infty \leq 1$, and so there exists $g \in L^1(G)$ satisfying $\phi = \hat{g}|_E$ and $\|g\|_1 \leq \varkappa$. Then we have

$$\|f\|_A = \sum_{\chi \in E} |\hat{f}(\chi)| = \sum_{\chi \in E} \phi(\chi)\hat{f}(\chi) = \sum_{\chi \in E} (g * f)^\wedge(\chi)$$

$$= g * f(0) \leq \|g * f\|_u \leq \|g\|_1 \|f\|_u \leq \varkappa \|f\|_u. \quad \square$$

Sidon sets satisfy various arithmetic conditions as we will show in Chapter 6. In the next theorem we give the only arithmetic condition needed for later results. Its elegant proof was shown to us by Sadahiro Saeki. Colin Graham also pointed out the relationship of this theorem to the corresponding result for Varopoulos' algebras; see Varopoulos [1967; section 6.3]. For applications of Theorem 1.4, see Corollary 4.6, Lemma 8.8, Corollary 5.14, and Theorem 6.14. An alternative proof and some brief historical remarks appear in 6.6 and 6.7.

1.4 THEOREM. _If_ E _is a Sidon set in_ X, _then_

$$\sup\{\min(|A|,|B|) : AB \subset E\} < \infty. \tag{1}$$

In particular, E _does not contain the product of two infinite sets_. [Here $|A|$ denotes the cardinal number of A and AB denotes the set $\{\chi\psi : \chi \in A,\ \psi \in B\}$.]

Proof. Assume that E is a Sidon set and that (1) fails. By 1.3(vii) there is a constant $\kappa > 0$ such that

$$\|f\|_A \leqq \kappa \|f\|_u \quad \text{for all} \quad f \in \mathrm{Trig}_E(G). \tag{2}$$

Let $n > 1$ be a fixed integer and select sets A and B such that $AB \subset E$ and $\min(|A|,|B|) \geqq n^3$. Let $\chi_1,\ldots,\chi_n$ be distinct elements of A and let $D_1 = \{\chi_1,\ldots,\chi_n\}$. Let ψ_1 be any element of $B \setminus D_1$. We inductively select $\psi_2,\ldots,\psi_n$ in B so that

$$\psi_k \notin C_k \equiv D_1 D_1^{-1} \cdot \{\psi_1,\ldots,\psi_{k-1}\} \cup D_1. \tag{3}$$

This selection is possible since $|C_k| \leqq n^2(n-1) + n < n^3 \leqq |B|$ for $k = 2,\ldots,n$. Let $D_2 = \{\psi_1,\ldots,\psi_n\}$ and $D = D_1 D_2$, and note that $D \subset AB \subset E$. [Note also that D_1 and D_2 are disjoint; we do not need this fact here, but will in 6.6.] It follows from (3) that the map $(\chi_j,\psi_k) \to \chi_j\psi_k$ is one-to-one on $D_1 \times D_2$ and hence $|D| = n^2$.

Now let $U = (u_{jk})_{j,k=1}^n$ be an $n \times n$ unitary matrix such that $|u_{jk}| = n^{-\frac{1}{2}}$ for all j and k. For example, we could let $u_{jk} = n^{-\frac{1}{2}}\exp(2\pi ijk/n)$ for all j and k. Since $D_1D_2 \subset E$, the trigonometric polynomial

$$g = \sum_{j,k=1}^{n} u_{jk}\chi_j\psi_k$$

belongs to $\mathrm{Trig}_E(G)$. Since the characters $\chi_j\psi_k$ are distinct we see that

$$\|g\|_A = n^2 n^{-\frac{1}{2}} = n^{3/2}. \tag{4}$$

To estimate $\|g\|_u$, we fix x in G and consider the vectors $\vec{\chi}(x) = (\overline{\chi_1(x)},\ldots,\overline{\chi_n(x)})$ and $\vec{\psi}(x) = (\psi_1(x),\ldots,\psi_n(x))$. Let $\langle\ ,\ \rangle$ signify the usual inner product in complex n-space and $\|\ \|$ the corresponding norm. Then we have

$$g(x) = \sum_{j=1}^{n}\left[\sum_{k=1}^{n} u_{jk}\psi_k(x)\right]\chi_j(x) = \langle U\vec{\psi}(x),\vec{\chi}(x)\rangle$$

and so Schwarz's inequality yields

$$|g(x)| \leqq \|U\vec{\psi}(x)\|\|\vec{\chi}(x)\| = \|\vec{\psi}(x)\|\cdot\|\vec{\chi}(x)\| = n^{\frac{1}{2}}n^{\frac{1}{2}} = n.$$

Since x is arbitrary, we conclude that

$$\|g\|_u \leqq n. \tag{5}$$

Applying (4), (2) and (5), we obtain

$$n^{3/2} = \|g\|_A \leqq \varkappa\|g\|_u \leqq \varkappa n .$$

Since n can be any integer greater than 1, we have a contradiction. □

Let Δ denote the space of maximal ideals for the commutative Banach algebra $M(G)$. Since $\mu \to \hat{\mu}(\chi)$ defines a member of Δ for each $\chi \in X$, we may regard X as a subset of Δ. The next theorem is also given in [DR; 5.2.5].

*1.5 THEOREM. <u>For</u> $E \subset X$, <u>the following are equivalent</u>:

(i) E <u>is a Sidon set</u>;

* An asterisk * attached to a result signifies that the result is not needed for the main results in the notes.

(ii) <u>the closure</u> $\overline{E}$ <u>of</u> E <u>in</u> Δ <u>is homeomorphic with the Stone-Čech compactification</u> βE <u>of</u> E;

(iii) <u>disjoint subsets of</u> E <u>have disjoint closures in</u> Δ.

<u>Proof</u>. Assume (i) holds and consider $\phi \in \ell^{\infty}(E)$. By 1.3(ii), we have $\phi = \hat{\mu}|_E$ for some $\mu \in M(G)$. Then the Gelfand transform $\hat{\mu}$ is a continuous function on Δ that extends ϕ. Hence every bounded function on E has a continuous extension to $\overline{E}$, so that (ii) holds.

The implication (ii) $\Rightarrow$ (iii) follows from general properties of βE, as follows. If E_1 and E_2 are disjoint subsets of E, then the characteristic function ϕ of E_1 on E has a continuous extension ϕ_o defined on $\overline{E}$. Then $\phi_o^{-1}(1)$ and $\phi_o^{-1}(0)$ are disjoint closed sets in Δ containing E_1 and E_2 respectively.

Suppose that (iii) holds and let $\overline{X}$ denote the closure of X in Δ. The family of functions $\{\hat{\mu}|_{\overline{X}} : \mu \in M(G)\}$ is clearly a subalgebra of $C(\overline{X})$ that separates points and does not vanish identically at any point. Moreover, this family is closed under conjugation because $(\nu^*)^\wedge$ coincides with $\overline{\hat{\nu}}$ on $\overline{X}$, even though we do not make this claim for Δ; see [R; 5.3]. Hence

$$\{\hat{\mu}|_{\overline{X}} : \mu \in M(G)\} \text{ is uniformly dense in } C(\overline{X}) \tag{1}$$

by the Stone-Weierstrass theorem. Now consider a function $\phi : E \to \{-1,1\}$. The disjoint sets $\phi^{-1}(1)$ and $\phi^{-1}(-1)$ have disjoint closures in Δ. Thus ϕ can be extended to a bounded continuous function ϕ_o on $\overline{X}$. Applying (1) to ϕ_o we find $\mu \in M(G)$ such that $|\hat{\mu}(\chi) - \phi(\chi)| < \frac{1}{2}$ for all $\chi \in E$. Thus E is a Sidon set by 1.3(iii). □

The next theorem shows that the space $C(G)$ in 1.3(vi) can be replaced by many smaller spaces of functions without destroying the validity of Theorem 1.3. The theorem is due to Edwards, Hewitt and Ross [1972b] who actually prove a somewhat more general result. The simple proof given here is due to Blei [1972a]. We begin by defining the smaller spaces.

*1.6 DEFINITION. For $1 < p < \infty$, let

$$A^p(G) = \{f \in C(G) : \hat{f} \in \ell^p(X)\}$$

and let

$$A^{1+}(G) = \cap\{A^p(G) : 1 < p < \infty\}.$$

For $w \in c_o(X)$, we define

$$A(G,w) = \{f \in C(G) : w\hat{f} \in \ell^1(X)\}.$$

Suppose that X is countable and let $\{X_n\}_{n=1}^{\infty}$ be an increasing sequence of finite subsets of X with $\cup_{n=1}^{\infty} X_n = X$. For any $f \in L^1(G)$, we define $s_n f = \Sigma_{\chi \in X_n} \hat{f}(\chi)\chi$; $U(G, \{X_n\})$ will denote the space of all functions f in $C(G)$ such that $\lim_{n\to\infty} \|s_n f - f\|_u = 0$. If the sequence $\{X_n\}$ is understood, we write $U(G)$ for $U(G, \{X_n\})$ so that

$$U(G) = \{f \in C(G) : s_n f \to f \text{ uniformly}\}.$$

The prototype for the sequences $\{X_n\}$ is given by $X_n = \{k \in \mathbb{Z} : |k| \leqq n\}$. In this case, $s_n f$ is the standard symmetric partial sum of the Fourier series of f and $U(\mathbb{T})$ is the space defined in [K; page 5].

In Edwards, Hewitt and Ross [1972b], these spaces are made into Banach or Frechet spaces and their conjugate spaces are determined.

*1.7 THEOREM. For $E \subset X$, the following are equivalent:

(i) E is a Sidon set;

(ii) $A_E^p(G) \subset A(G)$ for some $p > 1$;

(iii) $A_E^{1+}(G) \subset A(G)$;

(iv) $A_E(G,w) \subset A(G)$ for some $w \in c_o(X)$.

If X is countable, these statements are equivalent to:

(v) $U_E(G) \subset A(G)$.

Proof. It is clear from 1.3(vi) that (i) implies (ii) - (v). For the converses, we suppose that E is not a Sidon set and prove that each of (ii) - (v) also fails. If F is a finite subset of E, then $E \setminus F$ is not a Sidon set and so

$$\sup\{\|f\|_A/\|f\|_u : f \in \mathrm{Trig}_{E\setminus F}(G) \text{ and } f \neq 0\} = \infty \tag{1}$$

by 1.3(vii).

Now fix w in $c_o(X)$. A straightforward induction using (1) shows that there is a sequence $\{f_j\}_{j=1}^{\infty}$ in $\mathrm{Trig}_E(G)$ with the following properties:

$$\|f_j\|_A = 1/j, \tag{2}$$

$$\|f_j\|_u \leqq 2^{-j}, \tag{3}$$

and

$$\mathrm{Supp}(\hat{f}_j) \subset E \setminus [\bigcup_{k=1}^{j-1} \mathrm{Supp}(\hat{f}_k) \cup \{\chi \in X : |w(\chi)| \geqq 1/j\}] \tag{4}$$

for $j = 1,2,\ldots$. Let $f = \Sigma_{j=1}^{\infty} f_j$ and observe that $f \in C_E(G)$ by (3). Since the supports of the transforms $\hat{f}_j$ are pairwise disjoint, we have

$$\|\hat{f}\|_p^p = \sum_{j=1}^{\infty} \|\hat{f}_j\|_p^p \leqq \sum_{j=1}^{\infty} \|\hat{f}_j\|_1^p = \sum_{j=1}^{\infty} (\tfrac{1}{j})^p$$

for $p \geqq 1$. Hence $\|f\|_A = \|\hat{f}\|_1 = \infty$, so that $f \notin A(G)$. On the other hand, $\|\hat{f}\|_p < \infty$ for $p > 1$ and so f belongs to $A_E^p(G)$ for $p > 1$ and to $A_E^{1+}(G)$. It is also clear that f belongs to $A_E(G,w)$ since

$$\|w\hat{f}\|_1 = \sum_{j=1}^{\infty} \|w\hat{f}_j\|_1 \leqq \sum_{j=1}^{\infty} \frac{1}{j}\|\hat{f}_j\|_1 = \sum_{j=1}^{\infty} j^{-2} < \infty;$$

the first inequality holds because $|w(\chi)| < 1/j$ for χ in $\mathrm{Supp}(\hat{f}_j)$ by (4). We have shown that (ii) - (iv) fail.

Now suppose that X is countable and let $\{X_n\}$ be as in Definition 1.6. A simple induction shows that there is a sequence $\{f_j\}$ in $\mathrm{Trig}_E(G)$ and an increasing sequence $\{n_j\}$ in $\mathbb{Z}^+$ satisfying (2), (3) and

$$\mathrm{Supp}(\hat{f}_{j+1}) \subset E \setminus X_{n_j} \quad \text{and} \quad \mathrm{Supp}(\hat{f}_j) \subset X_{n_j} \tag{5}$$

for $j \geqq 1$. The supports of the transforms $\hat{f}_j$ are pairwise disjoint and so, as before, f belongs to $C_E(G) \setminus A(G)$. We claim that $f \in U(G)$, i.e. that $s_n f \to f$ uniformly. It is clear from (5) and (3) that

$$s_{n_j} f = \sum_{k=1}^{j} f_k \to f \quad \text{uniformly.} \tag{6}$$

Moreover, if $n_j \leqq n < n_{j+1}$, then

$$\|s_n f - s_{n_j} f\|_u \leqq \|s_n f - s_{n_j} f\|_A = \sum_{\chi \in X_n \setminus X_{n_j}} |\hat{f}(\chi)|$$

$$\leqq \sum_{\chi \in X_{n_{j+1}} \setminus X_{n_j}} |\hat{f}(\chi)| = \|f_{j+1}\|_A = \frac{1}{j+1} . \tag{7}$$

It follows from (6) and (7) that $\lim_{n\to\infty} \|s_n f - f\|_u = 0$. Hence f is in $U(G)$ and (v) fails. □

Corollary 5.14 and the discussion in 10.3 combine to give an example of a space $A_p(G)$ of functions, lying between $A(G)$

and $C(G)$, for which the inclusion $(A_p(G))_E \subset A(G)$ is <u>not</u> equivalent to the Sidonicity of E.

If $A(G)$ in 1.3(vi) or 1.3(v) is replaced by a larger function space, the validity of Theorem 1.3 seems invariably to be destroyed. For example, there are non-Sidon sets satisfying $C_E(G) \subset U(G)$ by 6.14. Non-Sidon sets satisfying $C_E(G) \subset A^p(G)$ for $p = 4/3$ are discussed in 10.6 and non-Sidon sets satisfying $L_E^\infty(G) \subset C(G)$ are discussed in 10.4.

Chapter 2

RIESZ PRODUCTS AND EXAMPLES OF SIDON SETS

A basic tool in the study of lacunarity is the Riesz product. The original Riesz products were trigonometric polynomials on $\mathbb{T} = [0,2\pi)$ of the form

$$\prod_{k=1}^{N}[1 + \cos(4^k t)];$$

these were studied by F. Riesz in 1918. In general groups, the possible appearance of characters of order 2 complicates matters. Before defining the general Riesz product we introduce some terminology. As in Chapter 1, G denotes a compact abelian group and X denotes its character group.

2.1 DEFINITIONS. A subset E of X is <u>symmetric</u> if $E = E^{-1}$. The set E is called <u>asymmetric</u> if $1 \notin E$ and

$$\chi \in E \text{ and } \chi^2 \neq 1 \text{ imply } \chi^{-1} \notin E. \tag{1}$$

A complex-valued function ϕ defined on a subset E of X is called <u>hermitian</u> if

$$\chi \in E,\ \chi^{-1} \in E \text{ imply } \phi(\chi^{-1}) = \overline{\phi(\chi)}. \tag{2}$$

In particular, ϕ must be real valued on elements of order 2. Note that a function ϕ on a set $E \subset X$ is hermitian if and only if ϕ can be extended to a hermitian function on the symmetric set $E \cup E^{-1}$.

A subscript h attached to a space of functions restricts the space to its hermitian elements.

Note that if μ is a real measure in $M(G)$, then $\hat{\mu}|_E$ is hermitian for all subsets E of X.

2.2 DEFINITION. A subset E of X is said to have the _Fatou-Zygmund property_ if there exists a constant $\kappa > 0$ so that to each ϕ in $\ell_h^\infty(E)$ there corresponds $\mu \in M^+(G)$ such that $\phi = \hat{\mu}|_E$ and $\|\mu\| \leqq \kappa\|\phi\|_\infty$.

2.3 REMARKS. If $\phi = \hat{\mu}|_E$ and $\mu \in M^+(G)$, then the hermitian extension of ϕ to $E \cup E^{-1}$ agrees with $\hat{\mu}$ on $E \cup E^{-1}$. Thus the set E has the Fatou-Zygmund property if and only if $E \cup E^{-1}$ has the Fatou-Zygmund property.

Every set E having the Fatou-Zygmund property is a Sidon set, as we now easily show. In view of the preceding paragraph, we may suppose that E is symmetric. Consider ϕ in $\ell^\infty(E)$. We define $\phi^\sim$ on E by $\phi^\sim(\chi) = \overline{\phi(\chi^{-1})}$. Thus ϕ is hermitian if and only if $\phi = \phi^\sim$. Since

$$\phi = \frac{\phi + \phi^\sim}{2} + i\,\frac{\phi - \phi^\sim}{2i},$$

we have

$$\phi = \phi_1 + i\phi_2 \quad \text{where } \phi_1, \phi_2 \text{ are hermitian,} \tag{1}$$

$$\|\phi_1\|_\infty \leqq \|\phi\|_\infty \quad \text{and} \quad \|\phi_2\|_\infty \leqq \|\phi\|_\infty. \tag{2}$$

The Fatou-Zygmund property implies that there are measures $\mu_1, \mu_2 \in M^+(G)$ satisfying $\hat{\mu}_j|_E = \phi_j$ and $\|\mu_j\| \leqq \kappa\|\phi_j\|_\infty$ for $j = 1,2$. If $\mu = \mu_1 + i\mu_2$, then $\phi = \hat{\mu}|_E$ and $\|\mu\| \leqq 2\kappa\|\phi\|_\infty$.

The remarkable fact is that the converse holds: Sidon sets without 1 satisfy the Fatou-Zygmund property. We prove this in Corollary 3.6.

Note that if $E \subset X$ has the Fatou-Zygmund property, then every bounded hermitian function on E is the restriction to E of some positive-definite function on X. In particular, it is easy to see that $1 \notin E$ whenever E has the Fatou-Zygmund property; see [HR; 32.4].

2.4 RIESZ PRODUCTS. Let F_0 be a finite asymmetric subset of X and let $F = F_0 \cup F_0^{-1}$. Let ϕ be any complex-valued function on F_0 such that $\phi(\chi)$ is real whenever $\chi^2 = 1$. Then ϕ is hermitian and can be extended to a hermitian function on F: simply define $\phi(\chi^{-1}) = \overline{\phi(\chi)}$ for $\chi \in F_0$. A Riesz product is a trigonometric polynomial p on G of the form

$$p = \prod_{\chi \in F_0} g_\chi \tag{1}$$

where each g_χ has the form

$$\left.\begin{aligned} g_\chi &= 1 + \phi(\chi)\chi + \phi(\chi^{-1})\chi^{-1} \quad \text{if} \quad \chi^2 \neq 1, \\ g_\chi &= 1 + \phi(\chi)\chi \quad \text{if} \quad \chi^2 = 1. \end{aligned}\right\} \tag{2}$$

The Riesz product p can, of course, also be written as

$$p = \sum_{\psi \in X} \hat{p}(\psi)\psi.$$

The coefficients $\hat{p}(\psi)$ can be determined by multiplying (1) out and collecting terms. We indicate one way that this can be done. We use the following notation: If $\{a_\alpha\}_{\alpha \in E}$ and $\{b_\alpha\}_{\alpha \in E}$ are families of complex numbers indexed by the finite set E, then

$$\prod_{\alpha \in E}(a_\alpha + b_\alpha) = \sum_{A \subset E} \prod_{\alpha \in A} a_\alpha \prod_{\alpha \in E \setminus A} b_\alpha, \tag{3}$$

where the sum is over all subsets A of E. Void products are taken to be 1. Now let

$$F_1 = \{\chi \in F_0 : \chi^2 \neq 1\} \quad \text{and} \quad F_2 = \{\chi \in F_0 : \chi^2 = 1\}.$$

Then we have

$$\begin{aligned} \prod_{\chi \in F_1} g_\chi &= \prod_{\chi \in F_1} [1 + \phi(\chi)\chi + \phi(\chi^{-1})\chi^{-1}] \\ &= \sum_{A \subset F_1} \prod_{\chi \in A}[\phi(\chi)\chi + \phi(\chi^{-1})\chi^{-1}] \prod_{\chi \in F_1 \setminus A} 1 \end{aligned}$$

$$= \sum_{A \subset F_1} \sum_{B \subset A} \prod_{\chi \in B} \phi(\chi)\chi \prod_{\chi \in A \setminus B} \phi(\chi^{-1})\chi^{-1}$$

$$= \sum_{B \subset A \subset F_1} \prod_{\chi \in B \cup (A \setminus B)^{-1}} \phi(\chi)\chi.$$

It is easy to see that $(B,A) \to B \cup (A \setminus B)^{-1}$ defines a one-to-one mapping of all pairs (B,A) satisfying $B \subset A \subset F_1$ onto the family of all asymmetric subsets of $F_1 \cup F_1^{-1}$. Thus we can write

$$\prod_{\chi \in F_1} g_\chi = \sum^{as}_{S_1 \subset F_1 \cup F_1^{-1}} \prod_{\chi \in S_1} \phi(\chi)\chi$$

where $\sum^{as}$ denotes the sum over all asymmetric sets S_1. We also have

$$\prod_{\chi \in F_2} g_\chi = \prod_{\chi \in F_2} [1 + \phi(\chi)\chi] = \sum_{S_2 \subset F_2} \prod_{\chi \in S_2} \phi(\chi)\chi.$$

Therefore

$$p = \prod_{\chi \in F_1 \cup F_2} g_\chi = \sum^{as}_{S_1 \subset F_1 \cup F_1^{-1}} \sum_{S_2 \subset F_2} \prod_{\chi \in S_1 \cup S_2} \phi(\chi)\chi.$$

Since $(S_1, S_2) \to S_1 \cup S_2$ defines a one-to-one mapping of all pairs (S_1, S_2) [where $S_1 \subset F_1 \cup F_1^{-1}$ is asymmetric and $S_2 \subset F_2$] onto the family of asymmetric subsets of $F = F_0 \cup F_0^{-1}$, we conclude that

$$p = \sum^{as}_{S \subset F} \prod_{\chi \in S} \phi(\chi)\chi. \tag{4}$$

We emphasize that $\sum^{as}_{S \subset F}$ signifies the sum over all asymmetric sets $S \subset F$. Now for $\psi \in X$, we have

$$\hat{p}(\psi) = \sum_{S} \prod_{\chi \in S} \phi(\chi), \tag{5}$$

where the sum is over all asymmetric sets $S \subset F$ for which

$$\prod_{\chi \in S} \chi = \psi. \tag{6}$$

The actual calculation of $\hat{p}(\psi)$ is normally a difficult task. If the set F_0 is suitably specialized, however, then the values of $\hat{p}(\psi)$ can be calculated exactly. We next define such sets.

2.5 DEFINITIONS. A subset E_0 of X is *independent* if $1 \notin E_0$ and for each finite subset F_0 of E_0 and integer-valued function m on F_0, the implication

$$\prod_{\chi \in F_0} \chi^{m(\chi)} = 1 \quad \text{implies} \quad \chi^{m(\chi)} = 1 \quad \text{for all} \quad \chi \tag{1}$$

holds. Similarly, E_0 is called *dissociate* if $1 \notin E_0$ and for each finite subset F_0 of E_0 and function m mapping F_0 into $\{-2,-1,0,1,2\}$, the implication (1) holds.

Independent sets are obviously dissociate, and dissociate sets are asymmetric. For examples and further discussion, see 2.8 - 2.19.

2.6 LEMMA. *Let* F_0 *be a finite dissociate set in* X *and let* $F = F_0 \cup F_0^{-1}$. *Let* ϕ *be a complex-valued function on* F_0 *such that*

$$\left.\begin{array}{l} |\phi(\chi)| \leqq \frac{1}{2} \ \text{for all} \ \chi \in F_0 \ \text{and} \\ \phi(\chi) \ \text{is real valued for} \ \chi = \chi^{-1} \in F_0. \end{array}\right\} \tag{1}$$

Extend ϕ *to be hermitian on* F. *The Riesz product* 2.4(1) *has these properties:*

$$p \geqq 0 \quad \text{and} \quad \|p\|_1 = \hat{p}(1) = 1, \tag{2}$$

$$\hat{p}\Big(\prod_{\chi \in S} \chi\Big) = \prod_{\chi \in S} \phi(\chi) \tag{3}$$

for asymmetric $S \subset F$, *and*

$$\hat{p}(\psi) = 0 \quad \text{for} \quad \psi \quad \text{in} \quad X \quad \text{not covered by (3).} \tag{4}$$

In particular,

$$\hat{p}(\chi) = \phi(\chi) \quad \text{for} \quad \chi \in F \tag{5}$$

and

$$|\hat{p}(\psi)| \leq \|\phi\|_\infty^2 \quad \text{for} \quad \psi \notin F \cup \{1\}. \tag{6}$$

Proof. The function p is defined by formulas 2.4(1) and 2.4(2). The hypotheses (1) imply that $g_\chi \geq 0$ for each χ, and therefore $p \geq 0$. This implies that $\|p\|_1 = \int_G p\, d\lambda = \hat{p}(1)$ and so (2) will be established as soon as (3) is [put $S = \emptyset$ in (3)]. The nice thing about dissociate sets is this: 2.5(1) implies that the characters

$$\psi_\varepsilon = \prod_{\chi \in F_0} \chi^{\varepsilon(\chi)},$$

where $\varepsilon : F_0 \to \{-1,0,1\}$ and $\varepsilon(\chi) \neq -1$ if $\chi^2 = 1$, are unique, i.e. $\varepsilon_1 \neq \varepsilon_2$ implies $\psi_{\varepsilon_1} \neq \psi_{\varepsilon_2}$. In set-theoretic terms, this means that the products $\prod_{\chi \in S} \chi$ in 2.4(4), where $S \subset F$ is asymmetric, are all distinct. Hence (3) holds. It is clear that (4) holds. □

*2.7 THEOREM. Let E_0 be a dissociate set in X and let $E = E_0 \cup E_0^{-1}$. If ϕ is a hermitian function on E where $\|\phi\|_\infty \leq \frac{1}{2}$, then there is a measure $\mu \in M^+(G)$ such that $\|\mu\| = 1$,

$$\hat{\mu}(\prod_{\chi \in S} \chi) = \prod_{\chi \in S} \phi(\chi) \tag{1}$$

for finite asymmetric $S \subset E$, and

$$\hat{\mu}(\psi) = 0 \quad \text{for } \psi \text{ in } X \text{ not covered by (1).} \tag{2}$$

In particular, E has the Fatou-Zygmund property with $\varkappa = 2$, and so E is a Sidon set.

---

* An asterisk * attached to a result signifies that the result is not needed for the main results in the notes.

<u>Proof</u>. For each finite symmetric subset F of E, Lemma 2.6 provides us with a Riesz product p_F satisfying

$$p_F \geqq 0 \quad \text{and} \quad \|p_F\| = 1, \tag{3}$$

$$\hat{p}_F\left(\prod_{\chi \in S} \chi\right) = \prod_{\chi \in S} \phi(\chi) \tag{4}$$

for asymmetric $S \subset F$, and

$$\hat{p}_F(\psi) = 0 \quad \text{for } \psi \text{ in } X \text{ not covered by (4).} \tag{5}$$

The finite symmetric subsets of E form a directed set under inclusion and so $\{p_F\}$ is a net in $M(G)$. By (3), each p_F belongs to $\{\mu \in M^+(G) : \|\mu\| = 1\}$, which is a norm bounded and weak-* closed subset of $M(G)$. By Alaoglu's theorem, the set $\{\mu \in M^+(G) : \|\mu\| = 1\}$ is compact in the weak-* topology. Hence the net $\{p_F\}$ has a subnet $\{p_\alpha\}$ that converges in the weak-* topology to a measure μ in $M^+(G)$ where $\|\mu\| = 1$. Since $\lim_\alpha \hat{p}_\alpha(\psi) = \hat{\mu}(\psi)$ for all $\psi \in X$, (1) and (2) are consequences of (4) and (5). □

Theorem 2.7 was proved by Hewitt and Zuckerman [1966] who coined the term "dissociate set"; see also [HR; Vol. II, 37.14]. A group X might have very few independent sets. For example, the nonvoid independent sets in $\mathbb{Z}$ have exactly one element. Our next result, taken from [HR; 37.18], shows that dissociate sets are more plentiful.

*2.8 THEOREM. <u>If X is infinite, then X contains an infinite dissociate subset. In fact, every infinite subset E of X contains a translate of such a set.</u>

<u>Proof</u>. <u>Case</u> 1: The set $\{\chi^2 : \chi \in E\}$ is infinite. We construct a dissociate subset of E by induction. Select any $\chi_1 \in E \setminus \{1\}$ and suppose $\chi_1, \chi_2, \ldots, \chi_n$ have been chosen so that

$\{\chi_1, \chi_2, \ldots, \chi_n\}$ is dissociate. The set

$$A_n = \{\chi_1^{\delta_1} \cdots \chi_n^{\delta_n} : \delta_j = 0, \pm 1, \pm 2, \pm 4\}$$

is finite, and so there is $\chi_{n+1} \in E$ such that $\chi_{n+1}^2 \notin A_n$. Assume that

$$\chi_1^{m_1} \cdots \chi_{n+1}^{m_{n+1}} = 1 \quad \text{where} \quad m_j = 0, \pm 1, \pm 2$$

and $m_{n+1} \neq 0$. We may assume that $m_{n+1} < 0$. Then

$$\chi_1^{m_1} \cdots \chi_n^{m_n} = \chi_{n+1} \quad \text{or} \quad \chi_{n+1}^2,$$

so that $\chi_{n+1}^2 = \chi_1^{m_1} \cdots \chi_n^{m_n}$ or $\chi_1^{2m_1} \cdots \chi_n^{2m_n}$. In either case, $\chi_{n+1}^2 \in A_n$, a contradiction. It follows that $\{\chi_n\}$ is an infinite dissociate subset of E.

Case 2: The set $\{\chi \in E : \chi^2 = 1\}$ is infinite. We may suppose that $\chi^2 = 1$ for all $\chi \in E$ and $1 \notin E$. Let χ_1 be any element in E and suppose $\chi_1, \ldots, \chi_n$ have been chosen so that $\{\chi_1, \ldots, \chi_n\}$ is independent. Let χ_{n+1} be any element in E not of the form $\chi_1^{m_1} \cdots \chi_n^{m_n}$ where $m_j = 0, 1$. Then $\{\chi_n\}$ is an infinite independent subset of E.

Case 3: The set $\{\chi^2 : \chi \in E\}$ is finite. In this case, there is an infinite subset E_1 of E such that $\chi^2 = \psi^2$ for all $\chi, \psi \in E_1$. Let χ_1 be any element in E_1. Then $\chi_1^{-1} E_1$ is an infinite set consisting of elements of order 2 and so Case 2 shows that $\chi_1^{-1} E_1$ contains an infinite dissociate [indeed independent] subset. □

*2.9 COROLLARY. *Every infinite subset of* X *contains an infinite subset that is a Sidon set.*

Proof. It is easy to check that the translate of a Sidon set is a Sidon set. Use, for example, property 1.3(vi). □

We next consider the classical lacunary sets that were studied so thoroughly by Sidon, Zygmund, Salem, and many others.

*2.10 DEFINITION. A set $E \subset \mathbb{Z}^+$ is called a Hadamard set if $E = \{n_k : k = 1,2,\ldots\}$ where

$$n_{k+1}/n_k \geqq q \quad \text{for all } k \tag{1}$$

where $q > 1$ is a constant.

*2.11 PROPOSITION. *If E is a Hadamard set with $q = 3$, then E is a dissociate set. Every Hadamard set is the finite union of such sets.*

Proof. Suppose that E is a Hadamard set satisfying 2.10(1) with $q = 3$. And suppose that for $n_1,\ldots,n_s \in \mathbb{Z}$, we have

$$\prod_{k=1}^{s}(e^{in_kx})^{m_k} = 1 \quad \text{where } m_k = 0, \pm 1, \pm 2.$$

Since this holds for all $x \in \mathbb{R}$, it follows that

$$\sum_{k=1}^{s} n_k m_k = 0.$$

We claim that $m_k = 0$ for all k. If not, we may assume that $m_s \neq 0$. Since $n_s \geqq 3n_{s-1} \geqq \cdots \geqq 3^k n_{s-k}$ for $0 \leqq k \leqq s-1$, we have

$$n_s \leqq |n_s m_s| = |-\sum_{k=1}^{s-1} n_k m_k| \leqq \sum_{k=1}^{s-1} n_k \cdot 2 = 2\sum_{k=1}^{s-1} n_{s-k}$$
$$\leqq 2\sum_{k=1}^{s-1} 3^{-k} n_s < 2n_s \sum_{k=1}^{\infty} 3^{-k} = n_s,$$

a contradiction. Thus $m_k = 0$ for all k.

Now consider any Hadamard set $E = \{n_k : k = 1,2,\ldots\}$. Choose $m \in \mathbb{Z}^+$ so that $q^m \geqq 3$. For $j = 1,2,\ldots,m$, let $E_j = \{n_{j+mk} : k = 0,1,2,\ldots\}$. Then each E_j is a Hadamard set satisfying 2.10(1) with $q = 3$ and $\bigcup_{j=1}^{m} E_j = E$. □

In 1926 Sidon proved that Hadamard sets are Sidon sets, i.e. that they satisfy 1.3(v). We can see this by citing Theorem 2.7 and Drury's Theorem 3.5, but this is very heavy-handed. A direct proof of a stronger result appears in 2.19.

*2.12 DISCUSSION. Stechkin [1956] and Rider [1966a] studied arithmetic conditions less stringent than 2.10(1) and 2.5(1). In order to describe them, we wish to define $R_s(E,\psi)$ for integers $s \geqq 0$, subsets E of X and characters ψ in X. In additive notation, $R_s(E,\psi)$ is usually defined to be the number of representations $\psi = \pm\chi_1 \pm\chi_2 \pm \cdots \pm\chi_s$ where $\chi_1,\ldots,\chi_s$ are in E; two representations are regarded as the same if the only difference is the ordering of the characters $\chi_1,\ldots,\chi_s$. See [R; page 124], Rider [1966a], Bonami [1970; page 351] and [DR; page 53]. The exact meaning of this definition does not usually matter, but it does if equation 2.15(13) below is to hold. This equation also appears in Rider [1966a; page 392, line 6] and in [DR; page 53, line -1]. The following concise definition of $R_s(E,\psi)$ is designed to make 2.15(13) valid.

*2.13 DEFINITIONS. For an integer $s \geqq 0$, subset E_0 of X and ψ in X, $R_s(E_0,\psi)$ denotes the number [possibly ∞] of asymmetric subsets S of $E = E_0 \cup E_0^{-1}$ satisfying

$$|S| = s \quad \text{and} \quad \prod_{\chi \in S} \chi = \psi. \tag{1}$$

Here $|S|$ denotes the cardinal number of S. Note that $R_s(E_0,\psi) = R_s(E,\psi)$ for all s and ψ.

We will call a set $E_0 \subset X$ a <u>Rider set</u> if there is a constant $B > 0$ such that $R_s(E_0,1) \leqq B^s$ for all s. We call a set a <u>Stechkin set</u> if it is a finite union of Rider sets.

Note that if E_0 is a dissociate set, then E_0 is a Rider set since $R_s(E_0,1) = 0$ for $s \geqq 1$ in this case; see the proof of 2.6.

*2.14 EXAMPLE. We give an example of a Rider set E in $\mathbb{Z}$ that is not a finite union of Hadamard sets; see Hewitt and Zuckerman [1959; page 8]. Let

$$E = \{2^n + 2^{m^2} : n,m \in \mathbb{Z}^+ \text{ and } (m-1)^2 \leqq n < m^2\}. \quad (1)$$

We show $R_s(E,0) \leqq 1$ for all $s \geqq 1$. Otherwise, for some $s > 1$ we could write

$$0 = \sum_{k=1}^{s} \varepsilon_k(2^{n_k} + 2^{m_k^2}) \quad (2)$$

where $n_1 < n_2 < \cdots < n_k$ and for each k, $\varepsilon_k = \pm 1$ and $(m_k - 1)^2 \leqq n_k < m_k^2$. Then 2^{n_1+1} divides every term appearing in (2) except the first one. This is a contradiction.

Now for any set $E \subset \mathbb{Z}^+$, let

$$N(E) = \sup\{|E \cap [y,2y]| : y \in \mathbb{R}^+\}.$$

If E is a Hadamard set satisfying 2.10(1) with $q = 3$, then each set $E \cap [y,2y]$ has at most one element. Hence by 2.11, we have $N(E) < \infty$ whenever E is a finite union of Hadamard sets. For the set E in (1), we have

$$|E \cap [2^{m^2}, 2\cdot 2^{m^2}]| = m^2 - (m-1)^2 = 2m - 1$$

for all $m \geqq 1$ and so E cannot be a finite union of Hadamard sets.

We will prove in 2.18 that if E is a Stechkin set, then $E \setminus \{1\}$ has a property similar to the Fatou-Zygmund property.

A corollary is that Stechkin sets are Sidon sets. See the comments after Corollary 2.19.

First we need to examine Riesz products more closely.

*2.15 RIESZ PRODUCTS REVISITED. Let E_0 be an asymmetric subset of X and let $E = E_0 \cup E_0^{-1}$. Let ϕ be a hermitian function defined on E. For each finite symmetric subset F of E, let $F_0 = F \cap E_0$ so that $F = F_0 \cup F_0^{-1}$ and F_0 is asymmetric, and let p_F denote the Riesz product 2.4(1). All we need to know from 2.4 is 2.4(4) - (6):

$$p_F = \sum_{S \subset F}^{as} \prod_{\chi \in S} \phi(\chi)\chi, \tag{1}$$

so that

$$\hat{p}_F(\psi) = \sum_S \prod_{\chi \in S} \phi(\chi) \quad \text{for} \quad \psi \in X, \tag{2}$$

where the sum is over all asymmetric sets $S \subset F$ for which

$$\psi = \prod_{\chi \in S} \chi. \tag{2'}$$

For $s \geqq 0$ and $\psi \in X$, we define

$$c_F^s(\psi) = \sum_S \prod_{\chi \in S} \phi(\chi), \tag{3}$$

where the sum is over all asymmetric sets $S \subset F$ for which

$$|S| = s \quad \text{and} \quad \psi = \prod_{\chi \in S} \chi. \tag{3'}$$

Formula (2) shows that

$$\hat{p}_F(\psi) = \sum_{s=0}^{|F|} c_F^s(\psi) \quad \text{for} \quad \psi \in X. \tag{4}$$

Observe that

$$c_F^0(1) = 1 \quad \text{and} \quad c_F^0(\psi) = 0 \quad \text{for} \quad \psi \neq 1, \tag{5}$$

$$c_F^1(\chi) = \phi(\chi) \quad \text{for} \quad \chi \in F, \tag{6}$$

and

$$c_F^1(\psi) = 0 \quad \text{for} \quad \psi \notin F. \tag{7}$$

Now suppose that there is some $\beta > 0$ such that $|\phi(\chi)| \leqq \beta$ for all $\chi \in E$. It is clear from (3) that

$$|c_F^s(\psi)| \leqq R_s(E,\psi)\beta^s \quad \text{for all } s \text{ and } \psi. \tag{8}$$

From (4) and (8) we get

$$|\hat{p}_F(\psi) - c_F^0(\psi) - c_F^1(\psi)| \leqq \sum_{s=2}^{\infty} R_s(E,\psi)\beta^s.$$

Hence by (5), (6) and (7), we have

$$|\hat{p}_F(1) - 1| \leqq \sum_{s=2}^{\infty} R_s(E,1)\beta^s, \tag{9}$$

$$|\hat{p}_F(\chi) - \phi(\chi)| \leqq \sum_{s=2}^{\infty} R_s(E,\chi)\beta^s \quad \text{for } \chi \in F, \tag{10}$$

and

$$|\hat{p}_F(\psi)| \leqq \sum_{s=2}^{\infty} R_s(E,\psi)\beta^s \quad \text{for } \psi \notin F \cup \{1\}. \tag{11}$$

Finally, suppose that $\phi(\chi) = \beta > 0$ for all $\chi \in E$. Then the definition of $R_s(F,\psi)$ and (3) show that $c_F^s(\psi) = R_s(F,\psi)\beta^s$ and so by (4)

$$\hat{p}_F(\psi) = \sum_{s=0}^{\infty} R_s(F,\psi)\beta^s \quad \text{for } \psi \in X, \tag{12}$$

where the sum in (12) has only a finite number of nonzero terms. Since

$$R_s(E,\psi) = \sup\{R_s(F,\psi) : F \text{ finite}, \ F \subset E\},$$

it is easy to see from (12) that

$$\sup\{\hat{p}_F(\psi) : F \text{ finite symmetric}, F \subset E\} = \sum_{s=0}^{\infty} R_s(E,\psi)\beta^s \tag{13}$$

for $\psi \in X$.

The next lemma is due to Rider [1966a].

*2.16 LEMMA. <u>If</u> E <u>is a Rider set and</u> $1 \notin E$, <u>then there is a constant</u> $B > 0$ <u>such that</u>

$$R_s(E,\psi) \leqq B^s \quad \underline{\text{for all}} \ \psi \in X \ \underline{\text{and}} \ s \geqq 0. \tag{1}$$

<u>Proof</u>. We may suppose that E is symmetric. By the definition, there is a constant $B_1 \geqq 1$ such that $R_s(E,1) \leqq B_1^s$ for all $s \geqq 0$. Let $\beta = \frac{1}{2B_1}$ and define $\phi(\chi) = \beta$ for $\chi \in E$. Consider the Riesz products p_F in 2.15. For each F and $\psi \in X$, we apply 2.15(12) to $1 \in X$ to obtain

$$|\hat{p}_F(\psi)| \leqq \|\hat{p}_F\|_u \leqq \|p_F\|_1 = \hat{p}_F(1) \leqq \sum_{s=0}^{\infty} R_s(E,1)\beta^s$$
$$\leqq \sum_{s=0}^{\infty} B_1^s \beta^s = \sum_{s=0}^{\infty} 2^{-s} = 2.$$

Now we apply 2.15(13) to ψ and obtain

$$\sum_{s=0}^{\infty} R_s(E,\psi)\beta^s = \sup\{\hat{p}_F(\psi) : F \subset E\} \leqq 2.$$

In particular, we have

$$R_s(E,\psi) \leqq 2\beta^{-s} = 2\cdot 2^s B_1^s \leqq (4B_1)^s$$

for all $s \geqq 0$. So (1) holds with $B = 4B_1$. □

*2.17 LEMMA. <u>Let</u> E <u>be a symmetric Rider set with</u> $1 \notin E$, <u>and let</u> $B \geqq 1$ <u>be so that 2.16(1) holds</u>. <u>If</u> $0 < \varepsilon < 1$, ϕ <u>is hermitian on</u> E <u>and</u> $\|\phi\|_\infty \leqq 1$, <u>then there exists</u> μ <u>in</u> $M^+(G)$ <u>so that</u>

$$\|\mu\| \leqq \varepsilon + 2B^2/\varepsilon, \tag{1}$$

$$|\hat{\mu}(\chi) - \phi(\chi)| \leqq \varepsilon \quad \underline{\text{for}} \quad \chi \in E, \tag{2}$$

<u>and</u>

$$|\hat{\mu}(\psi)| \leqq \varepsilon \quad \underline{\text{for}} \quad \psi \notin E \cup \{1\}. \tag{3}$$

<u>Proof</u>. Let $\beta = \varepsilon/(2B^2)$ and note that $0 < \beta \leqq \frac{1}{2}$. Let $\phi_1(\chi) = \beta\phi(\chi)$ for $\chi \in E$ and consider the Riesz products p_F in 2.15 defined by ϕ_1. Observe that for $\psi \in X$, we have

$$\sum_{s=2}^{\infty} R_s(E,\psi)\beta^s \leqq \sum_{s=2}^{\infty} (B\beta)^s = \sum_{s=2}^{\infty} \left(\frac{\varepsilon}{2B}\right)^s = \left(\frac{\varepsilon}{2B}\right)^2 \frac{1}{1 - \frac{\varepsilon}{2B}}$$

$$\leq \left(\frac{\varepsilon}{2B}\right)^2 \cdot 2 = \varepsilon \frac{\varepsilon}{2B^2} = \varepsilon\beta.$$

Applying this to inequalities (9), (10) and (11) in 2.15, we find

$$\|p_F\|_1 = \hat{p}_F(1) \leq 1 + \varepsilon\beta,$$

$$|\hat{p}_F(\chi) - \beta\phi(\chi)| \leq \varepsilon\beta \quad \text{for} \quad \chi \in F,$$

and

$$|\hat{p}_F(\psi)| \leq \varepsilon\beta \quad \text{for} \quad \psi \notin F \cup \{1\}.$$

By Alaoglu's theorem, the net $\{p_F\}$ in $M(G)$ has a weak-* cluster point $\nu \in M^+(G)$ such that

$$\|\nu\| \leq 1 + \varepsilon\beta,$$

$$|\hat{\nu}(\chi) - \beta\phi(\chi)| \leq \varepsilon\beta \quad \text{for} \quad \chi \in E,$$

and

$$|\hat{\nu}(\psi)| \leq \varepsilon\beta \quad \text{for} \quad \psi \notin E \cup \{1\}.$$

Clearly $\mu = \frac{1}{\beta}\nu$ satisfies (1), (2) and (3). □

*2.18 THEOREM. _Let_ E _be a symmetric Stechkin set with_ $1 \notin E$. _There is a constant_ C _with the following property. If_ $0 < \varepsilon < 1$, _if_ ϕ _is hermitian on_ E _and if_ $\|\phi\|_\infty \leq 1$, _then there exists_ $\mu \in M^+(G)$ _so that_

$$\|\mu\| \leq \varepsilon + C/\varepsilon, \tag{1}$$

$$|\hat{\mu}(\chi) - \phi(\chi)| \leq \varepsilon \quad \text{for} \quad \chi \in E, \tag{2}$$

and

$$|\hat{\mu}(\psi)| \leq \varepsilon \quad \text{for} \quad \psi \notin E \cup \{1\}. \tag{3}$$

Proof. We can write E as a disjoint union $\bigcup_{j=1}^m E_j$ where each E_j is a symmetric Rider set. Let $B \geq 1$ be so that 2.16(1) holds for each E_j. By 2.17, for each $j = 1,\ldots,m$ there is $\mu_j \in M^+(G)$ such that

$$\|\mu_j\| \leq \frac{\varepsilon}{m} + 2B^2m/\varepsilon, \tag{4}$$

$$|\hat{\mu}_j(\chi) - \phi(\chi)| \leq \frac{\varepsilon}{m} \quad \text{for} \quad \chi \in E_j, \tag{5}$$

and

$$|\hat{\mu}_j(\psi)| \leq \frac{\varepsilon}{m} \quad \text{for} \quad \psi \notin E_j \cup \{1\}. \tag{6}$$

Let $\mu = \mu_1 + \cdots + \mu_m$ and $C = 2B^2m^2$. Then (1) holds by (4). If $\chi \in E$, then χ belongs to exactly one of the sets E_k and so (5) and (6) yield

$$|\hat{\mu}(\chi) - \phi(\chi)| \leq |\hat{\mu}_k(\chi) - \phi(\chi)| + \sum_{j \neq k} |\hat{\mu}_j(\chi)|$$

$$\leq \frac{\varepsilon}{m} + (m-1)\frac{\varepsilon}{m} = \varepsilon.$$

Finally, if $\psi \notin E \cup \{1\}$, then (6) implies that

$$|\hat{\mu}(\psi)| \leq \sum_{j=1}^{m} |\hat{\mu}_j(\psi)| \leq m \cdot \frac{\varepsilon}{m} = \varepsilon. \quad \square$$

*2.19 COROLLARY. __Stechkin sets are Sidon sets.__

__Proof__. Verify 1.3(iii). □

Actually Stechkin sets without 1 satisfy the Fatou-Zygmund property. To see this directly we would need an analogue of Theorem 1.3, namely Theorem 7.2 [especially part (vii)] and Proposition 7.4. On the other hand, we will prove in 3.6 that all Sidon sets without 1 have the Fatou-Zygmund property, so we do not pursue this topic further at this point.

In his book [E; Vol. II, page 216], R. E. Edwards wondered whether every bounded function on an infinite Sidon set $E \subset \mathbb{Z}$ can be matched by a Fourier-Stieltjes transform $\hat{\mu}$ such that $\hat{\mu}|_{\mathbb{Z}\setminus E} \in c_0(\mathbb{Z}\setminus E)$. The following succinct answer was communicated to us by John Fournier.

*2.20 THEOREM. __Let__ E __be an infinite Sidon set in__ X __and consider__ $\phi \in \ell^\infty(E)$. __There exists a measure__ $\mu \in M(G)$ __such__ *that*

$$\hat{\mu}|_E = \phi \quad \textit{and} \quad \hat{\mu}|_{X\setminus E} \in c_o(X \setminus E) \tag{1}$$

if and only if

$$\phi \in c_o(E). \tag{2}$$

Proof. If ϕ belongs to $c_o(E)$, then $\phi = \hat{f}|_E$ for some $f \in L^1(G)$ by 1.3(iv). Since $\hat{f}$ belongs to $c_o(X)$, (1) holds.

To show that (1) implies (2), we assume the contrary. Then some μ in $M(G)$ satisfies

$$\hat{\mu}|_E \notin c_o(E) \quad \text{and} \quad \hat{\mu}|_{X\setminus E} \in c_o(X \setminus E). \tag{3}$$

For some $\delta > 0$, the set $\{\chi \in E : |\hat{\mu}(\chi)| \geqq \delta\}$ must be infinite. By 2.8 there is an infinite set

$$D \equiv \{\chi_n : n = 1,2,\ldots\} \subset \{\chi \in E : |\hat{\mu}(\chi)| \geqq \delta\} \tag{4}$$

such that ψD is dissociate for some $\psi \in X$. Since E is a Sidon set, 1.3(ii) provides us with $\mu_1 \in M(G)$ such that

$$\hat{\mu}_1(\chi_n) = 1 \quad \text{for all } n \text{ and } \hat{\mu}_1|_{E\setminus D} = 0. \tag{5}$$

By (3) - (5) the measure $\nu = \mu * \mu_1$ satisfies

$$|\hat{\nu}(\chi_n)| \geqq \delta \quad \text{for all } n \text{ and } \hat{\nu}|_{X\setminus D} \in c_o(X \setminus D). \tag{6}$$

Now let ν_o be a weak-* cluster point of the bounded sequence $\{\chi_n^{-1}\nu\}_{n=1}^{\infty}$ in $M(G)$. Since $(\chi_n^{-1}\nu)\hat{\ }(\chi) = \hat{\nu}(\chi\chi_n)$ for $\chi \in X$, it follows that $|\hat{\nu}_o(1)| \geqq \delta$. Moreover, we claim that

$$\hat{\nu}_o(\chi) = 0 \quad \text{for } \chi \neq 1. \tag{7}$$

If not, infinitely many characters $\chi\chi_n$ must belong to D since $\hat{\nu}|_{X\setminus D}$ is in $c_o(X \setminus D)$. So for infinitely many n, we can write $\chi\chi_n = \chi_{k_n}$. Since the χ_n's are distinct, there exist distinct integers n,m so that $\chi_m \notin \{\chi_n, \chi_{k_n}\}$ and

$$(\psi\chi_{k_n})(\psi\chi_n)^{-1}(\psi\chi_{k_m})^{-1}(\psi\chi_m) = \chi\chi^{-1} = 1.$$

This contradicts the fact that ψD is dissociate. Hence (7) holds and so ν_0 is a nonzero multiple of Haar measure λ on G. But by Helson's translation lemma [DR; page 64] (or [R; page 66]), ν_0 is singular with respect to λ. This contradiction completes the proof. □

Chapter 3

DRURY'S THEOREMS

In this chapter we prove two remarkable theorems due to S. W. Drury [1970], [1974]. In 1970 he proved that Sidon sets are uniformly Sidon sets (see Corollary 3.4). As a consequence, he showed that the union of two Sidon sets is again a Sidon set (Theorem 3.5). This settled a problem that had been outstanding for at least ten years.

In 1974 Drury proved that symmetric Sidon sets without 1 satisfy the Fatou-Zygmund property (Theorem 3.3); the converse is trivial and was proved in 2.3. This theorem makes it possible to improve many of the results concerning Fatou-Zygmund properties that were obtained by Edwards, Hewitt and Ross [1972c]. For the details, see Chapter 7.

The crux to both theorems is Drury's "convolution device" which we give in Lemma 3.2. We avoid duplicated effort by proving in Theorem 3.3 that symmetric Sidon sets without 1 have the Fatou-Zygmund property in a uniform way. Drury's earlier theorem that Sidon sets are uniformly Sidon sets is then obtained as a corollary. First we establish some notation that will be adhered to in this chapter.

3.1 NOTATION. For any real measure μ, μ^+ and μ^- are the positive and negative variations of μ, so that we have $\mu = \mu^+ - \mu^-$ and $|\mu| = \mu^+ + \mu^-$ thanks to the Jordan decomposition theorem. These notational conventions will <u>not</u> apply to Riesz products, however. Moreover, symbols like μ^{++} and μ^{--} have no predetermined meanings.

3.2 LEMMA. *Let* E *be a nonvoid finite (automatically Sidon) set in* X *with Sidon constant* κ. *Let* Ω *be a finite group of functions carrying* E *into* $\mathbb{T}$. *For each* $\omega \in \Omega$, *there exists a measure* μ_ω *in* $M(G)$ *with the following properties. If we write* $g_\chi(\omega) = \hat{\mu}_\omega(\chi)$ *for* $(\chi, \omega) \in X \times \Omega$, *then*

$$g_\chi(\omega) = \hat{\mu}_\omega(\chi) = \omega(\chi) \quad \text{for} \quad (\chi, \omega) \in E \times \Omega, \tag{1}$$

$$\|\mu_\omega\| \leqq \kappa^2 \quad \text{for} \quad \omega \in \Omega, \tag{2}$$

$$\|g_\chi\|_{A(\Omega)} \leqq \kappa^2 \quad \text{for} \quad \chi \in X. \tag{3}$$

If E *is symmetric and if all functions in* Ω *are hermitian, then each* μ_ω *can be taken to be a real measure. Moreover,* μ_ω *can be written as* $\mu_\omega^{++} - \mu_\omega^{--}$ *where* μ_ω^{++} *and* μ_ω^{--} *are nonnegative measures with the following properties. If we write* $\mu_\omega^* = \mu_\omega^{++} + \mu_\omega^{--}$ *and* $g_\chi^*(\omega) = (\mu_\omega^*)^\wedge(\chi)$ *for* $(\chi, \omega) \in E \times \Omega$, *then*

$$\|\mu_\omega^{\pm\pm}\| \leqq \kappa^2 \quad \text{for} \quad \omega \in \Omega, \tag{4}$$

$$\|g_\chi^*\|_{A(\Omega)} \leqq \kappa^2 \quad \text{for} \quad \chi \in X. \tag{5}$$

Proof. For each $\omega \in \Omega$ there is a measure $\nu_\omega \in M(G)$ such that

$$\hat{\nu}_\omega(\chi) = \omega(\chi) \quad \text{for all} \quad \chi \in E \quad \text{and} \quad \|\nu_\omega\| \leqq \kappa. \tag{6}$$

Now for each $\omega \in \Omega$, we define

$$\mu_\omega = \int_\Omega \nu_{\omega\alpha^{-1}} * \nu_\alpha \, d\alpha. \tag{7}$$

All integrals over Ω are with respect to normalized Haar measure and are really sums:

$$\int_{\Omega} h(\alpha)\, d\alpha = \frac{1}{|\Omega|} \sum_{\alpha\in\Omega} h(\alpha)$$

for functions h on Ω, where $|\Omega|$ denotes the cardinal number of Ω. To prove (1), we select $\chi \in E$ and compute:

$$\begin{aligned}\hat{\mu}_{\omega}(\chi) &= \int_{\Omega} (\nu_{\omega\alpha^{-1}})^{\wedge}(\chi)\hat{\nu}_{\alpha}(\chi)d\alpha = \int_{\Omega} (\omega\alpha^{-1})(\chi)\alpha(\chi)d\alpha \\ &= \int_{\Omega} \omega(\chi)d\alpha = \omega(\chi). \end{aligned} \tag{8}$$

To prove (2), we use (6) to write

$$\|\mu_{\omega}\| \leqq \int_{\Omega} \|\nu_{\omega\alpha^{-1}} * \nu_{\alpha}\|\, d\alpha \leqq \int_{\Omega} \|\nu_{\omega\alpha^{-1}}\|\,\|\nu_{\alpha}\|\, d\alpha \leqq \varkappa^2.$$

For $(\chi,\omega) \in X >\!\!< \Omega$, let $f_{\chi}(\omega) = \hat{\nu}_{\omega}(\chi)$. The first equality in (8) shows that

$$g_{\chi}(\omega) = \hat{\mu}_{\omega}(\chi) = \int_{\Omega} f_{\chi}(\omega\alpha^{-1})f_{\chi}(\alpha)d\alpha = f_{\chi} * f_{\chi}(\omega).$$

Hence we have

$$\|g_{\chi}\|_{A(\Omega)} = \|\hat{g}_{\chi}\|_1 = \|\hat{f}_{\chi}\hat{f}_{\chi}\|_1 = \|\hat{f}_{\chi}\|_2^2 = \|f_{\chi}\|_2^2 \leqq \|f_{\chi}\|_{\infty}^2.$$

Since $|f_{\chi}(\omega)| = |\hat{\nu}_{\omega}(\chi)| \leqq \|\nu_{\omega}\| \leqq \varkappa$ for all $\omega \in \Omega$, we see that $\|f_{\chi}\|_{\infty}^2 \leqq \varkappa^2$, so that (3) holds.

Now suppose that E is symmetric and that the functions in Ω are hermitian. For $\chi \in E$, we have by (6) that

$$\begin{aligned}(\mathrm{Re}\ \nu_{\omega})^{\wedge}(\chi) &= \tfrac{1}{2}[\hat{\nu}_{\omega}(\chi) + \hat{\bar{\nu}}_{\omega}(\chi)] = \tfrac{1}{2}[\hat{\nu}_{\omega}(\chi) + \overline{\hat{\nu}_{\omega}(\chi^{-1})}] \\ &= \tfrac{1}{2}[\omega(\chi) + \overline{\omega(\chi^{-1})}] = \omega(\chi);\end{aligned}$$

the last equality holds because ω is hermitian. This observation and the inequality $\|\mathrm{Re}\ \nu_{\omega}\| \leqq \varkappa$ show that ν_{ω} in (6) can be replaced by $\mathrm{Re}\ \nu_{\omega}$. I.e., we may assume that each ν_{ω} is a real measure.

Now for each $\omega \in \Omega$, we define

$$\mu_\omega^{++} = \int_\Omega (\nu^+_{\omega\alpha^{-1}} * \nu^+_\alpha + \nu^-_{\omega\alpha^{-1}} * \nu^-_\alpha)\,d\alpha,$$

$$\mu_\omega^{--} = \int_\Omega (\nu^+_{\omega\alpha^{-1}} * \nu^-_\alpha + \nu^-_{\omega\alpha^{-1}} * \nu^+_\alpha)\,d\alpha,$$

$$\mu_\omega^{*} = \mu_\omega^{++} + \mu_\omega^{--};$$

see 3.1. Clearly μ_ω^{++} and μ_ω^{--} are nonnegative measures. Also, we have

$$\mu_\omega^{++} - \mu_\omega^{--} = \int_\Omega [\nu^+_{\omega\alpha^{-1}} - \nu^-_{\omega\alpha^{-1}}] * [\nu^+_\alpha - \nu^-_\alpha]\,d\alpha$$

$$= \int_\Omega \nu_{\omega\alpha^{-1}} * \nu_\alpha \, d\alpha = \mu_\omega.$$

The proof of (4) is similar to that of (2). For example,

$$\|\mu_\omega^{++}\| \leqq \int_\Omega \| \nu^+_{\omega\alpha^{-1}} * \nu^+_\alpha + \nu^-_{\omega\alpha^{-1}} * \nu^-_\alpha \| \, d\alpha$$

$$\leqq \int_\Omega (\|\nu^+_{\omega\alpha^{-1}}\| + \|\nu^-_{\omega\alpha^{-1}}\|)(\|\nu^+_\alpha\| + \|\nu^-_\alpha\|)\,d\alpha$$

$$= \int_\Omega \|\nu_{\omega\alpha^{-1}}\| \, \|\nu_\alpha\| \, d\alpha \leqq \kappa^2.$$

The proof of (5) is similar to that of (3). In fact, for $(\chi,\omega) \in X \times \Omega$, let $h_\chi(\omega) = |\nu_\omega|\hat{\ }(\chi)$. Since

$$\mu_\omega^{*} = \int_\Omega [\nu^+_{\omega\alpha^{-1}} + \nu^-_{\omega\alpha^{-1}}] * [\nu^+_\alpha + \nu^-_\alpha]\,d\alpha = \int_\Omega |\nu_{\omega\alpha^{-1}}| * |\nu_\alpha|\,d\alpha,$$

we have

$$g_\chi^*(\omega) = (\mu_\omega^*)\hat{\ }(\chi) = \int_\Omega h_\chi(\omega\alpha^{-1}) h_\chi(\alpha)\,d\alpha = h_\chi * h_\chi(\omega)$$

and so

$$\|g_\chi^*\|_{A(\Omega)} = \|\hat{h}_\chi \hat{h}_\chi\|_1 = \|h_\chi\|_2^2 \leqq \|h_\chi\|_\infty^2 \leqq \kappa^2. \quad \square$$

3.3 DRURY'S THEOREM. <u>Let</u> E <u>be a symmetric Sidon set in</u> X <u>with Sidon constant</u> κ <u>such that</u> $1 \notin E$. <u>Given</u> $0 < \varepsilon \leqq 1$ <u>and a hermitian function</u> ϕ <u>on</u> E <u>such that</u> $\|\phi\|_\infty \leqq 1$, *there exists a nonnegative measure* $\mu \in M(G)$ *satisfying:*

$$\hat{\mu}|_E = \phi \quad \underline{\text{and}} \quad \|\mu\| \leq 32\kappa^4/\varepsilon, \tag{1}$$

and

$$|\hat{\mu}(\chi)| \leq \varepsilon \quad \underline{\text{for}} \quad \chi \notin E \cup \{1\}. \tag{2}$$

In particular, E *has the Fatou-Zygmund property.*

Proof. We may suppose that E is finite. For, suppose the result is known for all finite symmetric subsets F of E. Then there exist nonnegative measures μ_F such that

$$(\mu_F)^\wedge|_F = \phi|_F, \quad \|\mu_F\| \leq 32\kappa^4/\varepsilon,$$

and

$$|(\mu_F)^\wedge(\chi)| \leq \varepsilon \quad \text{for} \quad \chi \notin F \cup \{1\}.$$

The net $\{\mu_F : F$ finite symmetric, $F \subset E\}$ has a weak-* cluster point μ in $M^+(G)$ by Alaoglu's theorem, which satisfies (1) and (2).

Now let Ω consist of all hermitian functions ω mapping E into the four-element group $\mathbb{Z}(4) = \{1, i, -1, -i\}$. Let E_0 be an asymmetric subset of E such that $E_0 \cup E_0^{-1} = E$. Each ω in Ω is determined by its values on E_0, and any function ρ carrying E_0 into $\mathbb{Z}(4)$ has the form $\omega|_{E_0}$ for some $\omega \in \Omega$ provided only that $\rho(\chi) = \pm 1$ if $\chi^2 = 1$. Thus Ω is isomorphic to a product (indexed by E_0) of copies of $\mathbb{Z}(4)$ and $\mathbb{Z}(2) = \{1, -1\}$. Let H denote the set of all hermitian functions of E into the closed unit disc. Since the convex hull of the set $\mathbb{Z}(4)$ contains all complex numbers z such that $|z| \leq \cos(\pi/4)$, it follows that

$$H \subset \sec(\tfrac{\pi}{4})\ \mathrm{co}(\Omega) = \sqrt{2}\,\mathrm{co}(\Omega) \tag{3}$$

where $\mathrm{co}(\Omega)$ denotes the convex hull of Ω. To see this, it suffices to show that

$$H|_{E_0} \subset \sqrt{2}\,\mathrm{co}(\Omega|_{E_0}),$$

and this inclusion follows from previous remarks and the fact that if $A_1, A_2, \ldots, A_n$ are sets of complex numbers, then

$$\mathrm{co}(A_1) \times \cdots \times \mathrm{co}(A_n) = \mathrm{co}(A_1 \times \cdots \times A_n).$$

Let $\hat{\Omega}$ denote the character group of Ω, and let π be the natural injection of E_0 into $\hat{\Omega}$:

$$\pi_\chi(\omega) = \omega(\chi).$$

Viewing Ω as a product group, we see that each π_χ is simply a projection. Hence $\pi(E_0)$ is a dissociate set in $\hat{\Omega}$. Also $\mathrm{graph}(\pi) = \{(\chi, \pi_\chi) : \chi \in E_0\}$ is a dissociate set in $X \times \hat{\Omega}$, because

$$\prod_{\chi \in E_0} (\chi, \pi_\chi)^{m(\chi)} = 1 \quad \text{and} \quad m(\chi) \in \{-2,-1,0,1,2\}$$

imply that $\pi_\chi^{m(\chi)} = 1$ for $\chi \in E_0$ and this in turn implies that $m(\chi) = 0$ or else $\chi^2 = 1$.

Let $0 < \delta \leqq \frac{1}{2}$. We apply Lemma 2.6 twice, with $F_0 = \mathrm{graph}(\pi)$ and $\phi \equiv \delta$ on F_0 or $\phi \equiv -\delta$ on F_0, to obtain Riesz products p^+ and p^- on $G \times \Omega$ such that

$$p^{\pm} \geqq 0 \quad \text{and} \quad \|p^{\pm}\|_1 = 1, \tag{4}$$

$$(p^{\pm})\hat{}\,(\chi, \pi_\chi) = (p^{\pm})\hat{}\,(\chi^{-1}, \pi_\chi^{-1}) = \pm\delta \quad \text{for} \quad \chi \in E_0, \tag{5}$$

$$|(p^{\pm})\hat{}\,(\chi, \theta)| \leqq \delta^2 \quad \text{for} \quad (\chi,\theta) \notin \mathrm{graph}(\pi) \cup \mathrm{graph}(\pi)^{-1} \cup \{1\}. \tag{6}$$

Let $p^e = \frac{1}{2}(p^+ + p^-)$ and $p^o = \frac{1}{2}(p^+ - p^-)$; the letters e and o stand for even and odd. Then we have $p^+ = p^e + p^o$ and $p^- = p^e - p^o$. Since $(p^e)\hat{}\,(\chi, \pi_\chi) = (p^e)\hat{}\,(\chi^{-1}, \pi_\chi^{-1}) = 0$ for $\chi \in E_0$ by (5), we see from (6) that

$$|(p^e)^\wedge(\chi,\theta)| \leqq \delta^2 \quad \text{for} \quad (\chi,\theta) \neq 1. \tag{7}$$

Also

$$(p^o)^\wedge(\chi,\pi_\chi) = (p^o)^\wedge(\chi^{-1},\pi_\chi^{-1}) = \delta \quad \text{for} \quad \chi \in E_o, \tag{8}$$

and

$$|(p^o)^\wedge(\chi,\theta)| \leqq \delta^2 \quad \text{for} \quad \chi \notin E \cup \{1\}. \tag{9}$$

For $\omega \in \Omega$, we write $p_\omega^{\pm}$ for the nonnegative function on G defined by $p_\omega^{\pm}(x) = p^{\pm}(x,\omega)$. Observe that

$$\int_\Omega \|p_\omega^{\pm}\|_1 d\omega = \int_\Omega \int_G p^{\pm}(x,\omega)\,dx\,d\omega = \|p^{\pm}\|_1 = 1. \tag{10}$$

We now use the notation and results from Lemma 3.2. We define

$$\sigma_\omega = \int_\Omega (p^+_{\omega\alpha^{-1}} * \mu_\alpha^{++} + p^-_{\omega\alpha^{-1}} * \mu_\alpha^{--})\,d\alpha;$$

clearly σ_ω is in $M^+(G)$. Using 3.2(4) and (10), we obtain

$$\begin{aligned} \|\sigma_\omega\| &\leqq \int_\Omega (\|p^+_{\omega\alpha^{-1}}\|_1 \|\mu_\alpha^{++}\| + \|p^-_{\omega\alpha^{-1}}\|_1 \|\mu_\alpha^{--}\|)\,d\alpha \\ &\leqq \int_\Omega \varkappa^2(\|p^+_{\omega\alpha^{-1}}\|_1 + \|p^-_{\omega\alpha^{-1}}\|_1)\,d\alpha = 2\varkappa^2. \end{aligned} \tag{11}$$

A little algebraic manipulation allows us to rewrite σ_ω as

$$\sigma_\omega = \int_\Omega p^o_{\omega\alpha^{-1}} * \mu_\alpha\,d\alpha + \int_\Omega p^e_{\omega\alpha^{-1}} * \mu_\alpha^*\,d\alpha \equiv \sigma_\omega^o + \sigma_\omega^e.$$

Now we use 3.2(1) and (8) to compute

$$\begin{aligned} (\sigma_\omega^o)^\wedge(\chi) &= \int_\Omega (p^o_{\omega\alpha^{-1}})^\wedge(\chi)\hat{\mu}_\alpha(\chi)\,d\alpha \\ &= \int_\Omega \int_G p^o_{\omega\alpha^{-1}}(x)\overline{\chi(x)}\,dx\,\alpha(\chi)\,d\alpha \\ &= \int_\Omega \int_G p^o(x,\omega\alpha^{-1})\overline{\chi(x)}\,\pi_\chi(\alpha)\,dx\,d\alpha \\ &= \int_\Omega \int_G p^o(x,\omega\alpha^{-1})\overline{\chi(x)}\,\overline{\pi_\chi(\omega\alpha^{-1})}\,dx\,d\alpha \cdot \omega(\chi) \\ &= (p^o)^\wedge(\chi,\pi_\chi) \cdot \omega(\chi) = \delta\omega(\chi), \end{aligned} \tag{12}$$

for $\chi \in E_o$. Since $(\sigma_\omega^o)^\wedge$ and ω are hermitian, (12) holds for all $\chi \in E$.

Next we make some estimates for $(\sigma_\omega^o)^\wedge$ and $(\sigma_\omega^e)^\wedge$. Consider χ in X. Since $g_\chi(\omega) = \hat{\mu}_\omega(\chi)$, we have

$$(\sigma_\omega^o)^\wedge(\chi) = \int_\Omega (p^o_{\omega\alpha^{-1}})^\wedge(\chi) g_\chi(\alpha)d\alpha.$$

Since $g_\chi(\alpha) = \sum_{\theta\in\hat{\Omega}} \hat{g}_\chi(\theta)\theta(\alpha)$, we have

$$\begin{aligned}(\sigma_\omega^o)^\wedge(\chi) &= \int_\Omega \sum_{\theta\in\hat{\Omega}} (p^o_{\omega\alpha^{-1}})^\wedge(\chi)\hat{g}_\chi(\theta)\theta(\alpha)d\alpha \\ &= \int_\Omega \sum_{\theta\in\hat{\Omega}} \int_G p^o(x,\omega\alpha^{-1})\overline{\chi(x)}\hat{g}_\chi(\theta)\theta(\alpha)\,dx\,d\alpha \\ &= \sum_{\theta\in\hat{\Omega}} \int_\Omega\int_G p^o(x,\omega\alpha^{-1})\overline{\chi(x)\theta(\omega\alpha^{-1})}\,dx\,d\alpha\cdot\hat{g}_\chi(\theta)\theta(\omega) \\ &= \sum_{\theta\in\hat{\Omega}} (p^o)^\wedge(\chi,\theta)\hat{g}_\chi(\theta)\theta(\omega).\end{aligned}$$

Now if $\chi \notin E \cup \{1\}$, then (9) allows us to write

$$|(\sigma_\omega^o)^\wedge(\chi)| \leqq \delta^2 \sum_{\theta\in\hat{\Omega}} |\hat{g}_\chi(\theta)| = \delta^2 \|g_\chi\|_{A(\Omega)}$$

and so by 3.2(3),

$$|(\sigma_\omega^o)^\wedge(\chi)| \leqq \delta^2\kappa^2 \quad \text{for} \quad \chi \notin E \cup \{1\}. \tag{13}$$

In just the same way, one finds that

$$(\sigma_\omega^e)^\wedge(\chi) = \sum_{\theta\in\hat{\Omega}} (p^e)^\wedge(\chi,\theta)(g_\chi^*)^\wedge(\theta)\theta(\omega)$$

for $\chi \in X$. So if $\chi \neq 1$, then (7) and 3.2(5) lead to

$$|(\sigma_\omega^e)^\wedge(\chi)| \leqq \delta^2 \|g_\chi^*\|_{A(\Omega)} \leqq \delta^2\kappa^2 \quad \text{for} \quad \chi \neq 1. \tag{14}$$

We now reap the rewards of our labor. By (12) and (14), we have

$$|\hat{\sigma}_\omega(\chi) - \delta\omega(\chi)| \leqq \delta^2\kappa^2 \quad \text{for} \quad \chi \in E.$$

By (13) and (14), we have

$$|\hat{\sigma}_\omega(\chi)| \leqq 2\delta^2\kappa^2 \quad \text{for} \quad \chi \notin E \cup \{1\}.$$

Moreover, $\sigma_\omega \geqq 0$ and, as noted in (11), we have $\|\sigma_\omega\| \leqq 2\kappa^2$.

In other words, given $\omega \in \Omega$ there corresponds $\nu \in M^+(G)$ such that

$$|\hat{\nu}(\chi) - \omega(\chi)| \leqq \delta\kappa^2 \quad \text{for} \quad \chi \in E,$$
$$|\hat{\nu}(\chi)| \leqq 2\delta\kappa^2 \quad \text{for} \quad \chi \notin E \cup \{1\},$$

and

$$\|\nu\| \leqq 2\kappa^2/\delta.$$

These remarks also apply to convex combinations of functions in Ω. So by (3), if ϕ is hermitian on E and $\|\phi\|_\infty \leqq 1$, then some $\nu \in M^+(G)$ satisfies

$$|\hat{\nu}(\chi) - \phi(\chi)| \leqq \sqrt{2}\,\delta\,\kappa^2 \quad \text{for} \quad \chi \in E,$$
$$|\hat{\nu}(\chi)| \leqq 2\sqrt{2}\,\delta\,\kappa^2 \quad \text{for} \quad \chi \notin E \cup \{1\},$$

and

$$\|\nu\| \leqq 2\sqrt{2}\,\kappa^2/\delta.$$

Now select δ so that $4\sqrt{2}\,\delta\,\kappa^2 = \varepsilon$. It follows that if ϕ is a hermitian function on E, then some $\nu \in M^+(G)$ satisfies

$$|\hat{\nu}(\chi) - \phi(\chi)| \leqq \|\phi\|_\infty \frac{\varepsilon}{4} \quad \text{for} \quad \chi \in E,$$
$$|\hat{\nu}(\chi)| \leqq \|\phi\|_\infty \frac{\varepsilon}{2} \quad \text{for} \quad \chi \notin E \cup \{1\},$$

and

$$\|\nu\| \leqq \|\phi\|_\infty \frac{16\kappa^4}{\varepsilon}.$$

Finally, consider hermitian ϕ on E with $\|\phi\|_\infty \leqq 1$. We apply the foregoing inductively to ϕ, $\phi - \hat{\nu}_1$, etc. to obtain $\nu_1, \nu_2, \ldots$ in $M^+(G)$ such that

$$\left|\sum_{k=1}^{n} \hat{\nu}_k(\chi) - \phi(\chi)\right| \leqq \frac{\varepsilon}{4}\, 2^{-n+1} \quad \text{for} \quad \chi \in E,$$
$$|\hat{\nu}_k(\chi)| \leqq \frac{\varepsilon}{2}\, 2^{-k+1} \quad \text{for} \quad \chi \notin E \cup \{1\},$$

and

$$\|\nu_k\| \leqq \frac{16\kappa^4}{\varepsilon} \cdot 2^{-k+1}.$$

The measure $\mu = \sum_{k=1}^{\infty} \nu_k$ satisfies (1) and (2). □

Theorem 3.3 is due to Drury [1974]. The following corollary was proved by Drury [1970]. It tells us that Sidon sets are "uniformly approximable Sidon sets"; cf. Chaney [1969].

3.4 COROLLARY. _Let_ E _be a Sidon set in_ X _with Sidon constant_ κ, _and let_ $0 < \varepsilon \leqq 1$. _If_ $\phi \in \ell^{\infty}(E)$ _and_ $\|\phi\|_{\infty} \leqq 1$, _then there is a measure_ $\mu \in M(G)$ _such that_

$$\hat{\mu}|_E = \phi \quad \underline{\text{and}} \quad \|\mu\| \leqq 512\kappa^4/\varepsilon \tag{1}$$

and

$$|\hat{\mu}(\chi)| \leqq \varepsilon \quad \underline{\text{for}} \quad \chi \notin E. \tag{2}$$

Proof. Let $G_0 = G \rtimes \mathbb{T}$ so that its character group X_0 is $X \rtimes \mathbb{Z}$. Let $E_0 = E \rtimes \{1\}$. The idea of the proof is simple, but the details are tedious. We will show that $E_0 \cup E_0^{-1}$ is a Sidon set in X_0 with Sidon constant 2κ, then apply Theorem 3.3 to this set, and then pull the information back to E.

Consider ϕ_0 in $\ell^{\infty}(E_0 \cup E_0^{-1})$ and define $\phi(\chi) = \phi_0(\chi,1)$ and $\phi'(\chi) = \phi_0(\chi^{-1},-1)$ for $\chi \in E$. Since E is a Sidon set, there exist $\mu,\mu' \in M(G)$ so that $\hat{\mu}|_E = \phi$, $\hat{\mu}'|_E = \phi'$, $\|\mu\| \leqq \kappa\|\phi_0\|_{\infty}$, and $\|\mu'\| \leqq \kappa\|\phi_0\|_{\infty}$. Define μ_0 in $M(G_0)$ to be the measure corresponding to the linear functional

$$f \to \int_{\mathbb{T}}\int_G z\, f(x,z)\, d\mu(x)dz + \int_{\mathbb{T}}\int_G \bar{z}\, f(x,z)\, d\mu'(-x)dz$$

on $C(G_0)$, where dz signifies integration with respect to normalized Haar measure on $\mathbb{T}$. Clearly we have $\|\mu_0\| \leqq \|\mu\| + \|\mu'\| \leqq 2\kappa\|\phi_0\|_{\infty}$. Also we have $\hat{\mu}_0|_{E_0 \cup E_0^{-1}} = \phi_0$, since $\chi \in E$ implies

$$\hat{\mu}_0(\chi,1) = \int_{\mathbb{T}}\int_G z\,\overline{\chi(x)}\,\bar{z}\,d\mu(x)\,dz + \int_{\mathbb{T}}\int_G \bar{z}\,\overline{\chi(x)}\,\bar{z}\,d\mu'(-x)\,dz$$

$$= \hat{\mu}(\chi) = \phi(\chi) = \phi_0(\chi,1)$$

and

$$\hat{\mu}_0(\chi^{-1},-1) = \int_{\mathbb{T}}\int_G z\,\chi(x)\,z\,d\mu(x)\,dz + \int_{\mathbb{T}}\int_G \bar{z}\,\chi(x)\,z\,d\mu'(-x)\,dz$$

$$= \int_G \chi(x)d\mu'(-x) = \int_G \overline{\chi(x)}d\mu'(x)$$

$$= \hat{\mu}'(\chi) = \phi'(\chi) = \phi_0(\chi^{-1},-1);$$

note that $\int_{\mathbb{T}} \bar{z}^2 dz = \int_{\mathbb{T}} z^2 dz = 0$. Thus $E_0 \cup E_0^{-1}$ is a symmetric Sidon set with Sidon constant 2κ, and clearly $1 = (1,0)$ does not belong to $E_0 \cup E_0^{-1}$.

Now consider $0 < \varepsilon \leqq 1$ and $\phi \in \ell^{\infty}(E)$ where $\|\phi\|_{\infty} \leqq 1$. Define $\phi_0(\chi,1) = \phi(\chi)$ for $(\chi,1) \in E_0$ and extend ϕ_0 to a hermitian function on $E_0 \cup E_0^{-1}$. By Theorem 3.3, there is $\mu_0 \in M(G_0)$ satisfying

$$\hat{\mu}_0|_{E_0} = \phi_0|_{E_0} \quad \text{and} \quad \|\mu_0\| \leqq 32(2\kappa)^4/\varepsilon, \tag{3}$$

and

$$|\hat{\mu}_0(\psi)| \leqq \varepsilon \quad \text{for} \quad \psi \notin E_0 \cup E_0^{-1} \cup \{1\}.$$

In particular,

$$|\hat{\mu}_0(\chi,1)| \leqq \varepsilon \quad \text{for} \quad \chi \notin E, \tag{4}$$

since $(\chi,1)$ cannot possibly belong to $E_0^{-1} \cup \{1\}$. Let μ in $M(G)$ be the measure on G given by the linear functional $f \to \int_{G_0} f(x)\,\bar{z}\,d\mu_0(x,z)$ on $C(G)$. Then $\|\mu\| \leqq \|\mu_0\| \leqq 512\kappa^4/\varepsilon$. In addition, (3) and (4) imply (1) and (2), since

$$\hat{\mu}_0(\chi,1) = \int_{G_0} \overline{\chi(x)}\,\bar{z}\,d\mu_0(x,z) = \int_G \overline{\chi(x)}d\mu(x) = \hat{\mu}(\chi)$$

for $\chi \in X$. □

Drury [1970] used Corollary 3.4 to prove the next theorem.

3.5 DRURY'S THEOREM. The union of finitely many Sidon sets is a Sidon set.

Proof. It suffices to consider the union E of two Sidon sets E_1 and E_2. We may suppose that $E_1 \cap E_2 = \emptyset$. Now consider a function $\phi : E \to \{-1,1\}$. By Corollary 3.4 with $\varepsilon = \frac{1}{4}$, there are measures μ_1 and μ_2 in $M(G)$ such that

$$\hat{\mu}_1(\chi) = \phi(\chi) \quad \text{and} \quad |\hat{\mu}_2(\chi)| \leqq \tfrac{1}{4} \quad \text{for} \quad \chi \in E_1,$$

and

$$\hat{\mu}_2(\chi) = \phi(\chi) \quad \text{and} \quad |\hat{\mu}_1(\chi)| \leqq \tfrac{1}{4} \quad \text{for} \quad \chi \in E_2.$$

If $\mu = \mu_1 + \mu_2$, then $|\hat{\mu}(\chi) - \phi(\chi)| \leqq \frac{1}{4}$ for $\chi \in E$, and so 1.3(iii) holds. Hence E is a Sidon set by Theorem 1.3. □

3.6 COROLLARY. Let E be a Sidon set in X such that $1 \notin E$. Then E has the Fatou-Zygmund property.

Proof. It is easy to see that E^{-1} is a Sidon set. Hence by Theorem 3.5, $E \cup E^{-1}$ is a Sidon set and so Theorem 3.3 shows that $E \cup E^{-1}$ has the Fatou-Zygmund property. As noted in 2.3, E must also have the Fatou-Zygmund property. □

When Corollary 3.4 became available, S. Hartman and B. B. Wells independently showed that a set E is a Sidon set if and only if every bounded function on E can be matched by the Fourier-Stieltjes transform of a continuous measure. We will give this result in Corollary 4.8.

3.7 REMARKS. Some comments about the auxiliary groups used in the proofs of 3.3 and 3.4 seem in order. The proof of Theorem 3.3 remains unchanged, except for the values of some constants, if Ω is taken to be all hermitian functions mapping E into $\mathbb{Z}(n)$ for an even integer $n \geqq 4$. We simply used the smallest such n. A proof can even be made if $\mathbb{Z}(n)$ is replaced by $\mathbb{T}$, but in this case Lemma 3.2 needs to be extended to infinite Ω and in its proof the measures ν_ω need to be selected in a measurable way. In the proof of Corollary 3.4, $\mathbb{Z}$ can be replaced by any discrete abelian group containing an element that is not of order 2. Our choice of $\mathbb{Z}$ will be convenient in the proof of Corollary 8.18. A direct proof of Corollary 3.4 can be given using (1) - (3) of Lemma 3.2. See Drury [1972].

Drury [1974a] modifies the techniques of this chapter to show that the Sidonicity of a set $E \subset X$ depends only on the set of functions $n : E \to \{-1,0,1\}$ that have finite support and satisfy $\prod_{\chi \in E} \chi^{n(\chi)} = 1$ and $\sum_{\chi \in E} n(\chi) = 0$. Note that Rider sets and Stechkin sets are, by definition, determined by the set of functions $n : E \to \{-1,0,1\}$ that have finite support and satisfy $\prod_{\chi \in E} \chi^{n(\chi)} = 1$. Stechkin sets are the only known Sidon sets; see Corollary 2.19 and Remark 9.2.

Chapter 4

THE HARTMAN-WELLS THEOREM

By Theorem 1.3, a subset E of X is a Sidon set if and only if each ϕ in $\ell^{\infty}(E)$ has the form $\hat{\mu}|_E$ for some μ in $M(G)$. If $1 \notin E$, Corollary 3.6 (to Drury's Theorem 3.3 and 3.5) shows that the measure μ can be taken to be nonnegative provided ϕ is hermitian. In this chapter we prove that μ can even be taken to be a continuous measure. This result is due to S. Hartman and, independently, to B. B. Wells; see 4.7 and 4.8. In our preliminary work 4.1 - 4.5, G need not be compact; as usual X will denote its character group.

4.1 WIENER'S LEMMA. _Let_ G _be an LCA (locally compact abelian) group and consider an open basis of neighborhoods_ V _of_ 0. _For each_ V, _there is a nonnegative continuous positive-definite function_ h_V _such that_ $\hat{h}_V \in L^1(X)$, $h_V(0) = 1$ _and_ $\mathrm{Supp}(h_V) \subset V$. _For each_ $\mu \in M(G)$,

$$\sum_{x \in G} |\mu(\{x\})|^2 = \lim_V \int_X \hat{h}_V\, |\hat{\mu}|^2 d\theta, \tag{1}$$

where θ _denotes Haar measure on_ X.

Proof. The following proof is given in [HR; 41.18.a] and [R; 5.6.9]. For each V, let V' be a compact symmetric neighborhood of 0 such that $V' + V' \subset V$, and let $h_V = \lambda(V')^{-1}\xi_{V'} * \xi_{V'}$ where ξ_S denotes the characteristic function of S. Then each h_V is a continuous nonnegative positive-definite function on G such that $\hat{h}_V \in L^1(X)$, $h_V(0) = 1$ and $\mathrm{Supp}(h_V) \subset V$.

To prove (1), let $\nu = \mu * \mu^*$, so that $\hat{\nu} = |\hat{\mu}|^2$. Applying Fubini's theorem and the inversion theorem [HR; 33.10] to h_V, we obtain

$$\int_X |\hat{\mu}|^2 \hat{h}_V \, d\theta = \int_X \hat{\nu} \hat{h}_V \, d\theta = \int_G h_V(-x) d\nu(x).$$

Simple arguments now show that

$$\lim_V \int_X \hat{h}_V |\hat{\mu}|^2 d\theta = \lim_V \int_G h_V(-x) d\nu(x) = \nu(\{0\})$$

$$= \mu * \mu^*(\{0\}) = \int_G \overline{\mu(\{x\})} d\mu(x) = \sum_{x \in G} |\mu(\{x\})|^2. \quad \square$$

4.2 COROLLARY. *If* μ *is a continuous measure on an LCA group* G, *then*

$$\inf\{|\hat{\mu}(\chi)| : \chi \in X\} = 0. \tag{1}$$

Proof. If (1) fails, we may assume that $|\hat{\mu}(\chi)| \geqq 1$ for all $\chi \in X$. Since $\int_X \hat{h}_V d\theta = h_V(0) = 1$ for all V, we can use 4.1(1) to write

$$1 = \lim_V \int_X \hat{h}_V d\theta \leqq \lim_V \int_X \hat{h}_V |\hat{\mu}|^2 d\theta = \sum_{x \in G} |\mu(\{x\})|^2 = 0,$$

a clear contradiction. $\square$

For a measure μ on an LCA group G, the discrete and continuous parts of μ will be denoted by μ_d and μ_c, respectively.

4.3 THEOREM. *Let* G *be a nondiscrete LCA group with character group* X, *and suppose that* $E \subset X$ *satisfies*

$$\sup\{\min(|A|, |B|) : AB \subset E\} < \infty. \tag{1}$$

For $\mu \in M(G)$, *we have*

$$\hat{\mu}_d(X) \subset \hat{\mu}(X \setminus E)^-. \tag{2}$$

[*Note that each side of (2) refers to a set of complex numbers.*]

Proof. Let χ_o in X and $\varepsilon > 0$ be fixed. We first show that there is a sequence $\{\chi_n\}_{n=1}^{\infty}$ in X of distinct characters satisfying

$$|\hat{\mu}_c(\chi_o \chi_n \chi_m^{-1})| < \varepsilon \quad \text{for} \quad n > m \geqq 1. \tag{3}$$

Select χ_1 arbitrarily and suppose distinct $\chi_1, \ldots, \chi_{n-1}$ have been chosen to satisfy (3). Let g_n be a function in $L^1(G)$ such that $\hat{g}_n(\chi_m) = \varepsilon$ for $1 \leqq m < n$, and consider the following function on X:

$$\phi_n(\chi) = \sum_{m=1}^{n-1} |\hat{\mu}_c(\chi_o \chi \chi_m^{-1})|^2 + |\hat{g}_n(\chi)|^2.$$

Since G is not discrete, ϕ_n is the Fourier-Stieltjes transform of a continuous measure, namely

$$\sum_{m=1}^{n-1} (\chi_o^{-1} \chi_m \mu_c) * (\chi_o^{-1} \chi_m \mu_c)^* + g_n * g_n^*.$$

By Corollary 4.2, some χ_n in X satisfies $|\phi_n(\chi_n)| < \varepsilon^2$. Then $|\hat{\mu}_c(\chi_o \chi_n \chi_m^{-1})| < \varepsilon$ for $m = 1, 2, \ldots, n-1$ and also $\chi_n \neq \chi_m$ for $1 \leqq m < n$ since $|\hat{g}_n(\chi_n)| < \varepsilon$ while $\hat{g}_n(\chi_m) = \varepsilon$. This completes the inductive step and establishes (3).

Now we write μ_d as $\sum_{j=1}^{\infty} a_j \, \delta_{-x_j}$ where a_j is a complex number, δ_{-x_j} denotes the point mass at $-x_j$, and $\sum_{j=1}^{\infty} |a_j| < \infty$. Note that $\hat{\mu}_d(\chi) = \sum_{j=1}^{\infty} a_j \, \chi(x_j)$ for $\chi \in X$. Choose $N \in \mathbb{Z}^+$ so that $\sum_{j=N+1}^{\infty} |a_j| < \varepsilon$. Passing to subsequences, if necessary, we may assume that $\lim_{n\to\infty} \chi_n(x_j)$ exists for each $j = 1, 2, \ldots, N$, say

$$\lim_{n\to\infty} \chi_n(x_j) = c_j \quad \text{for} \quad j = 1, 2, \ldots, N. \tag{4}$$

Now for $n > m \geqq 1$, (3) shows that

$$\begin{aligned}
|\hat{\mu}_d(\chi_o) &- \hat{\mu}(\chi_o\chi_n\chi_m^{-1})| \\
&\leqq |\hat{\mu}_d(\chi_o) - \hat{\mu}_d(\chi_o\chi_n\chi_m^{-1})| + |\hat{\mu}_c(\chi_o\chi_n\chi_m^{-1})| \\
&< |\sum_{j=1}^{\infty} a_j[\chi_o(x_j) - \chi_o\chi_n\chi_m^{-1}(x_j)]| + \varepsilon \\
&\leqq \sum_{j=1}^{N} |a_j|\cdot|1 - \chi_n(x_j)\chi_m^{-1}(x_j)| + 2\sum_{j=N+1}^{\infty} |a_j| + \varepsilon \\
&< 3\varepsilon + \sum_{j=1}^{N} |a_j|\cdot|1 - \chi_n(x_j)\chi_m^{-1}(x_j)|.
\end{aligned}$$

Now (4) shows that there exists $M \in \mathbb{Z}^+$ so that $n > m \geqq M$ implies $\Sigma_{j=1}^{N}|a_j|\cdot|1 - \chi_n(x_j)\chi_m^{-1}(x_j)| < \varepsilon$ and hence

$$|\hat{\mu}_d(\chi_o) - \hat{\mu}(\chi_o\chi_n\chi_m^{-1})| < 4\varepsilon.$$

Since χ_o and ε are arbitrary, to prove (2) it suffices to show that $\chi_o\chi_n\chi_m^{-1} \in X \setminus E$ for some $n > m \geqq M$. Assume that $\chi_o\chi_n\chi_m^{-1} \in E$ for $n > m \geqq M$. If $A_m = \{\chi_{2m}, \ldots, \chi_{3m-1}\}\cdot\chi_o$ and $B_m = \{\chi_m^{-1}, \ldots, \chi_{2m-1}^{-1}\}$, then $|A_m| = |B_m| = m$ and $A_mB_m \subset E$. Since m can be any integer $\geqq M$, this contradicts (1). □

4.4 COROLLARY. *If G is a nondiscrete LCA group and $\mu \in M(G)$, then $\hat{\mu}_d(X) \subset \hat{\mu}(X)^-$.*

4.5 COROLLARY. *If G is a nondiscrete LCA group and $\mu \in M(G)$, then $\|\hat{\mu}_d\|_\infty \leqq \|\hat{\mu}\|_\infty$.*

4.6 COROLLARY. *If G is a compact infinite abelian group and $E \subset X$ is a Sidon set, then*

$$\|\hat{\mu}_d\|_\infty \leqq \sup\{|\hat{\mu}(\chi)| : \chi \in X \setminus E\}. \tag{1}$$

Proof. The set E satisfies 4.3(1) by Theorem 1.4. □

Corollary 4.4 is due to Glicksberg and Wik [1972]. Corollary 4.5 is a consequence of a theorem of Eberlein [1955] concerning weakly almost periodic functions; this result is explicitly stated in Dunkl and Ramirez [1972a]. Corollary 4.6 is due to Wells [1973], who used it to prove Theorem 4.7 below. Theorem 4.7 was proved independently by Hartman [1972b]. Henceforth G is a compact abelian group.

4.7 HARTMAN-WELLS THEOREM. *Let E be a Sidon set in X with $1 \notin E$. A constant κ' exists so that given ϕ in $\ell_h^\infty(E)$, there is a continuous nonnegative measure $\mu \in M(G)$ satisfying $\hat{\mu}|_E = \phi$ and $\|\mu\| \leq \kappa' \|\phi\|_\infty$.*

Proof. Since $E \cup E^{-1}$ is a Sidon set by Drury's Theorem 3.5, we may suppose that E itself is symmetric. Now let κ denote the Sidon constant of E. By Drury's Theorem 3.3, there is $\mu \in M^+(G)$ such that $\hat{\mu}|_E = 1$, $\|\mu\| \leq 64\kappa^4$, and

$$|\hat{\mu}(\chi)| \leq \tfrac{1}{4} \quad \text{for} \quad \chi \notin E \cup \{1\}.$$

By Corollary 4.6 applied to $E \cup \{1\}$, we have $\|\hat{\mu}_d\|_\infty \leq \frac{1}{4}$ and so

$$|\hat{\mu}_c(\chi)| \geq \tfrac{1}{2} \quad \text{for} \quad \chi \in E. \tag{1}$$

Consider a function ϕ in $\ell_h^\infty(E)$. Since μ_c is a nonnegative measure, $\hat{\mu}_c$ is hermitian. Hence $\phi/(\hat{\mu}_c|_E)$ is hermitian on E and by (1) it is bounded by $2\|\phi\|_\infty$. By 3.3 again, there is $\nu \in M^+(G)$ such that $\hat{\nu}|_E = \phi/(\hat{\mu}_c|_E)$ and $\|\nu\| \leq 32\,\kappa^4 \cdot 2\|\phi\|_\infty$. Now $\nu*\mu_c$ is continuous and nonnegative, $(\nu*\mu_c)\hat{}\,|_E = \phi$, and

$$\|\nu*\mu_c\| \leq \|\nu\| \cdot \|\mu_c\| \leq \|\nu\| \cdot \|\mu\| \leq (64)^2\,\kappa^8\,\|\phi\|_\infty.$$

Let $\kappa' = (64)^2\,\kappa^8$. □

4.8 COROLLARY [HARTMAN-WELLS]. *Let E be a Sidon set in X. Given $\phi \in \ell^\infty(E)$ there is a continuous measure $\mu \in M(G)$ such that $\hat{\mu}|_E = \phi$.*

Proof. Same as for 4.7 except use Corollary 3.4 in place of Theorem 3.3. □

In Hartman's proof of Theorem 4.7 the next theorem served the role of a lemma.

*4.9 THEOREM. Let G be a compact infinite abelian group. If E is a Sidon set in X, then $X \setminus E$ is dense in the Bohr compactification bX of X.

Proof. If not, the Fourier algebra $A(bX)$ contains a nonzero function g that vanishes on $X \setminus E$. The character group of bX is G_d, i.e. G with the discrete topology. Moreover, given $x \in G$ the corresponding character $\hat{x}$ on bX satisfies $\hat{x}(\chi) = \chi(x)$ for $\chi \in X$. Since $M(G_d) = \ell^1(G_d)$, it follows that $g = \hat{\mu}$ for some $\mu \in M(G_d)$. Then μ is a discrete measure in $M(G)$ and $\hat{\mu}(\chi) = \int_G \bar{\chi}\, d\mu = g(\chi)$ for $\chi \in X$ whether G is regarded as discrete or not. Hence by 4.6(1) we have

$$\|\hat{\mu}|_X\|_\infty \leqq \sup\{|\hat{\mu}(\chi)| : \chi \in X \setminus E\} = 0;$$

thus $\mu = 0$ and $g = 0$, a contradiction. □

Let E be an infinite Sidon set in X and let $\bar{E}$ denote its closure in bX. It is not known whether $\bar{E}$ must be proper in bX. It may be that $\bar{E}$ must have measure zero in bX. In fact, $\bar{E}$ might always be a Helson set in bX.

Chapter 5

$\Lambda(q)$ SETS AND SIDON SETS

The main result in this chapter is Theorem 5.8 which asserts that Sidon sets are $\Lambda(q)$ sets for $1 < q < \infty$. We will also show that the converse fails: there are non-Sidon sets that are $\Lambda(q)$ sets for all q, $1 < q < \infty$; see Corollary 5.14.

First we state Hölder's inequality in a useful and easily remembered form.

5.1 HÖLDER'S INEQUALITY. *Let* μ *be a positive measure on some space, let* $1 \leq p < r < q \leq \infty$, *and write*

$$\frac{1}{r} = \alpha\cdot\frac{1}{p} + (1-\alpha)\cdot\frac{1}{q} \quad \text{where} \quad 0 < \alpha < 1. \tag{1}$$

If $f \in L^p(\mu) \cap L^q(\mu)$, *then* $f \in L^r(\mu)$ *and*

$$\|f\|_r \leq \|f\|_p^{\alpha} \|f\|_q^{1-\alpha}. \tag{2}$$

Proof. Suppose $q < \infty$. By (1), we have

$$1 = \frac{\alpha r}{p} + \frac{(1-\alpha)r}{q}.$$

Hence for $f \in L^p(\mu) \cap L^q(\mu)$, the standard Hölder inequality gives us

$$\int |f|^r d\mu = \int |f|^{\alpha r} |f|^{(1-\alpha)r} d\mu$$

$$\leq \left(\int |f|^{\alpha r \cdot \frac{p}{\alpha r}} d\mu\right)^{\alpha r/p} \left(\int |f|^{(1-\alpha)r\cdot\frac{q}{(1-\alpha)r}} d\mu\right)^{(1-\alpha)r/q}$$

$$= \left(\int |f|^p d\mu\right)^{\alpha r/p} \left(\int |f|^q d\mu\right)^{(1-\alpha)r/q} = \|f\|_p^{\alpha r} \|f\|_q^{(1-\alpha)r}$$

Hence $f \in L^r(\mu)$ and (2) holds. A similar and simpler argument works if $q = \infty$. $\square$

5.2 DEFINITION. A set $E \subset X$ is a $\Lambda(q)$ *set*, $1 < q < \infty$, if a constant $\varkappa_q$ exists such that

$$\|f\|_q \leqq \varkappa_q \|f\|_1 \quad \text{for all} \quad f \in \mathrm{Trig}_E(G). \tag{1}$$

5.3 THEOREM. *Consider* $E \subset X$ *and* $1 < q < \infty$. *The following statements are equivalent*:

(i) E *is a* $\Lambda(q)$ *set*,

(ii) $\mu \in M_E(G)$ *implies* $\mu = f\lambda$ *where* $f \in L^q(G)$,

(iii) $f \in L^1_E(G)$ *implies* $f \in L^q(G)$.

For $q > 2$, *these properties are equivalent to*

(iv) $f \in L^2_E(G)$ *implies* $f \in L^q(G)$,

(v) *for some constant* η_q,

$$\|f\|_q \leqq \eta_q \|f\|_2 \quad \textit{for all} \quad f \in \mathrm{Trig}_E(G), \tag{1}$$

(vi) *for the same constant* η_q,

$$f \in L^{q'}(G) \ \textit{implies} \ \hat{f}|_E \in \ell^2(E) \ \textit{and} \ \|\hat{f}|_E\|_2 \leqq \eta_q \|f\|_{q'}; \tag{2}$$

here q' *is the conjugate exponent of* q.

Proof. *(i) ⇒ (ii)*. Let $\{h_\alpha\}$ be an approximate unit for $L^1(G)$ consisting of trigonometric polynomials satisfying $\|h_\alpha\|_1 \leqq 1$ for all α; see [HR; 28.53]. If $\mu \in M_E(G)$, then each $\mu * h_\alpha$ belongs to $\mathrm{Trig}_E(G)$ and so 5.2(1) implies that

$$\|\mu * h_\alpha\|_q \leqq \varkappa \|\mu * h_\alpha\|_1 \leqq \varkappa \|\mu\| \cdot \|h_\alpha\|_1 \leqq \varkappa \|\mu\|$$

for all α. By Alaoglu's theorem, the net $\{\mu * h_\alpha\}$ in $L^q(G)$ has a weak-* cluster point f in $L^q(G)$. Since $\lim_\alpha \hat{h}_\alpha(\chi) = 1$ for all $\chi \in X$, it follows that $\hat{f} = \hat{\mu}$. Uniqueness of the Fourier-Stieltjes transform now tells us that $\mu = f\lambda$.

(ii) ⇒ (iii). Obvious.

<u>(iii) ⇒ (i)</u>. By (iii), the map $f \to f$ of $L^q_E(G)$ into $L^1_E(G)$ is an onto map, and it is continuous since $\|f\|_1 \leqq \|f\|_q$. It is easy to see that $L^q_E(G)$ and $L^1_E(G)$ are closed subspaces of $L^q(G)$ and $L^1(G)$, respectively, and hence are Banach spaces. The open mapping theorem provides $\kappa_q > 0$ such that $\|f\|_q \leqq \kappa_q \|f\|_1$ for all $f \in L^1_E(G)$. Thus E is a Λ(q) set.

Henceforth we assume that $2 < q < \infty$.

<u>(iii) ⇒ (iv)</u>. Obvious.

<u>(iv) ⇒ (v)</u>. Imitate the proof of (iii) ⇒ (i) with L^2 in place of L^1 throughout.

<u>(v) ⇒ (i)</u>. For some α, $0 < \alpha < 1$, we have $\frac{1}{2} = \alpha \cdot \frac{1}{1} + (1-\alpha)\cdot\frac{1}{q}$. By (1) and Hölder's inequality 5.1(2), we have

$$\|f\|_q \leqq \eta_q \|f\|_2 \leqq \eta_q \|f\|_1^{\alpha} \|f\|_q^{1-\alpha}$$

and hence $\|f\|_q^{\alpha} \leqq \eta_q \|f\|_1^{\alpha}$ for $f \in \mathrm{Trig}_E(G)$. Thus 5.2(1) holds with $\kappa_q = \eta_q^{1/\alpha}$.

<u>(v) ⇒ (vi)</u>. It is easy to see that (1) holds for all f in $L^2_E(G)$. We define $T : \ell^2(E) \to L^q(G)$ by $T(\phi) = \check{\phi}$, where $\check{\phi}$ denotes the inverse L^2 transform. Then T is clearly a bounded linear map and $\|T\| \leqq \eta_q$. We consider the adjoint map $T^* : L^{q'}(G) \to \ell^2(E)$. For $f \in L^2(G)$ and $\phi \in \ell^2(E)$, we have

$$\langle \phi, T^*(f) \rangle = \int_G \check{\phi}\,\bar{f}\, d\lambda = \langle \phi, \hat{f} \rangle$$

and so $T^*(f) = \hat{f}|_E$. Since $L^2(G)$ is dense in $L^{q'}(G)$, we have $T^*(f) = \hat{f}|_E$ for $f \in L^{q'}(G)$. Hence for $f \in L^{q'}(G)$ we have $\hat{f}|_E \in \ell^2(E)$ and

$$\|\hat{f}|_E\|_2 = \|T^*(f)\|_2 \leq \|T^*\| \, \|f\|_{q'} = \|T\| \, \|f\|_{q'} \leq \eta_q \|f\|_{q'}.$$

Thus (2) holds.

<u>(vi) ⇒ (v)</u>. Here we simply reverse the preceding argument. By (2), $T(f) = \hat{f}|_E$ defines a bounded linear map of $L^{q'}(G)$ into $\ell^2(E)$ such that $\|T\| \leq \eta_q$. The adjoint map T^* : $\ell^2(E) \to L^q(G)$ satisfies $\langle T^*(\phi), g\rangle = \langle \check{\phi}, g\rangle$ for $g \in L^2(G)$ and hence for $g = \chi \in X$. Therefore $T^*(\phi) = \check{\phi}$ for all $\phi \in \ell^2(E)$. In particular, for $f \in \mathrm{Trig}_E(G)$, we have $T^*(\hat{f}|_E) = f$ and so

$$\|f\|_q = \|T^*(\hat{f}|_E)\|_q \leq \|T^*\| \, \|\hat{f}|_E\|_2 = \|T\| \, \|\hat{f}\|_2 \leq \eta_q \|f\|_2.$$

This establishes (v). □

Other properties of sets E equivalent to those listed in Theorem 5.3 can be found in [HR; 37.9]. See also 10.3. We now work towards our main result: Sidon sets are $\Lambda(q)$ sets.

5.4 DEFINITION (compare 2.13). For $E \subset X$, $\psi \in X$ and $s \in \mathbb{Z}^+$, we will denote by $r_s(E,\psi)$ the cardinal number of the set of all s-tuples $(\chi_1, \ldots, \chi_s) \in E^s$ satisfying

$$\prod_{j=1}^{s} \chi_j = \psi.$$

5.5 THEOREM. <u>Consider</u> $E \subset X$ <u>and suppose that there exist</u> $N \in \mathbb{Z}^+$ <u>and</u> $s \geq 2$ <u>such that</u>

$$r_s(E,\psi) \leq N \quad \text{for all} \quad \psi \in X. \tag{1}$$

<u>Then</u> E <u>is a</u> $\Lambda(2s)$ <u>set and</u>

$$\|f\|_{2s} \leq N^{1/2s} \|f\|_2 \quad \text{for all} \quad f \in \mathrm{Trig}_E(G). \tag{2}$$

<u>Proof</u>. Consider f in $\mathrm{Trig}_E(G)$ and let $g = f^s$. Then $\hat{g}$ is an s-fold convolution of $\hat{f}$ and so

$$\hat{g}(\psi) = \sum_{\chi_1 \cdots \chi_s = \psi} \hat{f}(\chi_1) \cdots \hat{f}(\chi_s) \quad \text{for} \quad \psi \in X.$$

By Hölder's inequality we have

$$|\hat{g}(\psi)|^2 \leq [\sum_{\chi_1\cdots\chi_s=\psi} |\hat{f}(\chi_1)|^2\cdots|\hat{f}(\chi_s)|^2][\sum_{\chi_1\cdots\chi_s=\psi} 1^2]$$

$$= r_s(E,\psi)[\sum_{\chi_1\cdots\chi_s=\psi} |\hat{f}(\chi_1)|^2\cdots|\hat{f}(\chi_s)|^2]$$

for $\psi \in X$ and so

$$\|f\|_{2s}^{2s} = \|g\|_2^2 \leq N \sum_{\psi\in X} \sum_{\chi_1\cdots\chi_s=\psi} |\hat{f}(\chi_1)|^2\cdots|\hat{f}(\chi_s)|^2$$

$$= N \sum_{(\chi_1,\ldots,\chi_s)\in E^s} |\hat{f}(\chi_1)|^2\cdots|\hat{f}(\chi_s)|^2$$

$$= N[\sum_{\chi\in E} |\hat{f}(\chi)|^2]^s \leq N\|f\|_2^{2s}.$$

This proves (2). □

Theorem 5.5 appears in Bonami [1970; page 356].

5.6 NOTATION. As in Chapter 3, we need auxiliary groups. Let I be an index set and let Ω be the group of all functions ω mapping I into $\mathbb{T}$. Then $\Omega = \mathbb{T}^I$, a compact abelian group with the product topology. For each $i \in I$, π_i will denote the i-th projection: $\pi_i(\omega) = \omega(i)$ or ω_i for $\omega \in \Omega$. We will denote the set $\{\pi_i : i \in I\}$ by I also. Then I is an independent set in the character group of Ω. Hence I is a dissociate set and a Sidon set by 2.7. We next use Theorem 5.5 to show that I is a $\Lambda(q)$ set for $1 < q < \infty$. This will be used to show that all Sidon sets are $\Lambda(q)$ sets, which is an instance where I-spectral polynomials on Ω can be used in place of the classical Rademacher functions (see [E; § 14.1]).

5.7 THEOREM. <u>Let</u> I <u>and</u> Ω <u>be as in</u> <u>5.6</u>. <u>For</u> f <u>in</u> $\mathrm{Trig}_I(\Omega)$, <u>we have</u>

$$\|f\|_{2s} \leqq (s!)^{1/2s}\|f\|_2 \quad \underline{\text{for}} \quad s \in \mathbb{Z}^+, \tag{1}$$

$$\|f\|_2 \leqq \sqrt{2}\,\|f\|_1, \tag{2}$$

<u>and</u>

$$\|f\|_q \leqq \sqrt{q}\,\|f\|_2 \leqq \sqrt{2q}\,\|f\|_1 \quad \underline{\text{for}} \quad 2 < q < \infty. \tag{3}$$

<u>Proof</u>. Let $s \geqq 2$ and consider ψ in the character group of Ω. Since I is independent, two s-tuples $(\pi_1,\ldots,\pi_s)$ and $(\pi_1',\ldots,\pi_s')$ in I^s will satisfy $\prod_{j=1}^{s}\pi_j = \prod_{j=1}^{s}\pi_j' = \psi$ if and only if the s-tuples are permuations of each other. Hence for each ψ, we have $r_s(I,\psi) = s!$ or $r_s(I,\psi) = 0$. Therefore (1) holds by Theorem 5.5 with $N = s!$.

To verify (2) and (3) we consider a fixed f in $\mathrm{Trig}_I(\Omega)$. Then (1) with $s = 2$ shows that $\|f\|_4^4 \leqq 2\|f\|_2^4$. Since $\frac{1}{2} = \frac{2}{3}\cdot\frac{1}{4} + \frac{1}{3}\cdot\frac{1}{1}$, Hölder's inequality tells us that $\|f\|_2 \leqq \|f\|_4^{2/3}\|f\|_1^{1/3}$. Hence we have

$$\|f\|_2^6 \leqq \|f\|_4^4\|f\|_1^2 \leqq 2\|f\|_2^4\|f\|_1^2,$$

so that $\|f\|_2^2 \leqq 2\|f\|_1^2$ and (2) holds.

Since $s! \leqq s^s$ for $s \in \mathbb{Z}^+$, (1) implies $\|f\|_{2s} \leqq \sqrt{s}\,\|f\|_2$. Given $2 < q < \infty$, select an integer $s \geqq 2$ satisfying $2s - 2 \leqq q \leqq 2s$. Then $s \leqq q$ and so

$$\|f\|_q \leqq \|f\|_{2s} \leqq \sqrt{s}\,\|f\|_2 \leqq \sqrt{q}\,\|f\|_2.$$

Since $\sqrt{q}\,\|f\|_2 \leqq \sqrt{2q}\,\|f\|_1$ by (2), (3) is established. □

5.8 THEOREM. *Let $E \subset X$ be a Sidon set with Sidon constant $\varkappa$. Then E is a $\Lambda(q)$ set for $1 < q < \infty$. In fact, for $f \in \mathrm{Trig}_E(G)$, we have*

$$\|f\|_q \leqq \varkappa\sqrt{q}\,\|f\|_2 \quad \text{for } q > 2, \tag{1}$$

and

$$\|f\|_2 \leqq \varkappa\sqrt{2}\,\|f\|_1. \tag{2}$$

[Note that (1) establishes 5.3(v) with $\eta_q = \varkappa\sqrt{q}$.]

Proof. By arguments seen in the proof of Theorem 5.7, it clearly suffices to prove

$$\|f\|_{2s}^{2s} \leqq \varkappa^{2s}\, s!\, \|f\|_2^{2s} \quad \text{for } s \in \mathbb{Z}^+, \tag{3}$$

and to prove (2) for $f \in \mathrm{Trig}_E(G)$. To this end, fix a nonzero $f \in \mathrm{Trig}_E(G)$ and write $f = \sum_{\chi\in E} \hat{f}(\chi)\chi$. Let $I = E$, and let $\Omega = \mathbb{T}^I$ as in 5.6 and 5.7. We need the following auxiliary function on $G \times \Omega$:

$$F(x,\omega) = F_x(\omega) = F_\omega(x) = \sum_{\chi\in E} \hat{f}(\chi)\omega(\chi)\chi(x) = \sum_{\chi\in E} \hat{f}(\chi)\chi(x)\pi_\chi(\omega).$$

It is easy to see that $F \in \mathrm{Trig}(G \times \Omega)$, $F_x \in \mathrm{Trig}(\Omega)$ for $x \in G$, and $F_\omega \in \mathrm{Trig}(G)$ for $\omega \in \Omega$. Since E is a Sidon set, to each $\omega \in \Omega$ there corresponds $\mu_\omega \in M(G)$ such that $\hat{\mu}_\omega(\chi) = \overline{\omega(\chi)}$ for $\chi \in E$ and $\|\mu_\omega\| \leqq \varkappa$. We next show that

$$F_\omega * \mu_\omega = f \quad\text{and}\quad f * \mu_\omega = F_{\omega^{-1}} \quad \text{for all } \omega \in \Omega. \tag{4}$$

Since the functions in (4) are in $\mathrm{Trig}_E(G)$, it suffices to check that the corresponding Fourier transforms are equal on E. But $\chi \in E$ implies

$$(F_\omega)^\wedge(\chi)\hat{\mu}_\omega(\chi) = \hat{f}(\chi)\omega(\chi)\overline{\omega(\chi)} = \hat{f}(\chi)$$

and

$$\hat{f}(\chi)\hat{\mu}_\omega(\chi) = \hat{f}(\chi)\overline{\omega(\chi)} = \hat{f}(\chi)\omega^{-1}(\chi) = (F_{\omega^{-1}})^\wedge(\chi).$$

Now for $s \in \mathbb{Z}^+$, we have

$$\|f\|_{2s} = \|F_\omega * \mu_\omega\|_{2s} \leqq \|F_\omega\|_{2s} \|\mu_\omega\| \leqq \varkappa \|F_\omega\|_{2s} \tag{5}$$

and

$$\|F_{\omega^{-1}}\|_1 = \|f * \mu_\omega\|_1 \leqq \|f\|_1 \|\mu_\omega\| \leqq \varkappa \|f\|_1 . \tag{6}$$

Applying (5), 5.7(1) to the functions F_x, and familiar facts, we obtain

$$\begin{aligned}
\|f\|_{2s}^{2s} &= \int_\Omega \|f\|_{2s}^{2s} d\omega \leqq \varkappa^{2s} \int_\Omega \|F_\omega\|_{2s}^{2s} d\omega = \varkappa^{2s} \int_\Omega \int_G |F_\omega(x)|^{2s}\, dx\, d\omega \\
&= \varkappa^{2s} \int_G \int_\Omega |F_x(\omega)|^{2s} d\omega\, dx \leqq \varkappa^{2s} \int_G s! \|F_x\|_2^{2s} dx \\
&= \varkappa^{2s} s! \int_G \|F_x^\wedge\|_2^{2s} dx = \varkappa^{2s} s! \int_G [\sum_{\chi \in E} |\hat{f}(\chi)\chi(x)|^2]^s dx \\
&= \varkappa^{2s} s! \|\hat{f}\|_2^{2s} = \varkappa^{2s} s! \|f\|_2^{2s};
\end{aligned}$$

this proves (3). In a similar way, we establish (2) using 5.7(2) and (6):

$$\begin{aligned}
\|f\|_2 &= \|\hat{f}\|_2 = [\sum_{\chi \in E} |\hat{f}(\chi)|^2]^{\frac{1}{2}} = \int_G [\sum_{\chi \in E} |\hat{f}(\chi)\chi(x)|^2]^{\frac{1}{2}} dx \\
&= \int_G \|F_x^\wedge\|_2 dx = \int_G \|F_x\|_2 dx \leqq \sqrt{2} \int_G \|F_x\|_1 dx \\
&= \sqrt{2} \int_G \int_\Omega |F_x(\omega)|\, d\omega\, dx = \sqrt{2} \int_\Omega \int_G |F_\omega(x)|\, dx\, d\omega \\
&= \sqrt{2} \int_\Omega \|F_\omega\|_1 d\omega \leqq \sqrt{2}\, \varkappa \int_\Omega \|f\|_1 d\omega = \varkappa \sqrt{2} \|f\|_1 . \quad \square
\end{aligned}$$

5.9 REMARK. As a companion to Theorem 5.8, we will prove in Corollary 5.14 that every infinite group X contains non-Sidon sets that are Λ(q) sets for all finite $q > 1$. To do this, we will first show that some relatively fat sets are Λ(q) sets. Items 5.10 - 5.13 are taken from Bonami [1970; pp. 356-360]. Here is a generalization of Theorem 5.7.

*5.10 THEOREM. Let $\Omega = \mathbb{T}^I$ be as in 5.6. For $k \in \mathbb{Z}^+$, the set I^k of all products of k not-necessarily-distinct projections π_i is a $\Lambda(q)$ set for all finite $q > 1$. Furthermore, for $f \in \mathrm{Trig}_{I^k}(\Omega)$, we have

$$\|f\|_{2s} \leqq s^{k/2}\|f\|_2 \quad \text{for} \quad s \in \mathbb{Z}^+, \tag{1}$$

$$\|f\|_q \leqq q^{k/2}\|f\|_2 \quad \text{for} \quad 2 < q < \infty, \tag{2}$$

and

$$\|f\|_2 \leqq 2^k\|f\|_1. \tag{3}$$

Proof. Inequalities (2) and (3) are deduced from (1) just as (2) and (3) in Theorem 5.7 were deduced from 5.7(1). By Theorem 5.5 with $N = s^{ks}$, to prove (1) it suffices to show that

$$r_s(I^k,\psi) \leqq s^{ks} \tag{4}$$

for characters ψ in the character group of Ω. So we consider an s-tuple $(\psi_1,\ldots,\psi_s)$ of elements from I^k such that $\psi = \prod_{j=1}^s \psi_j$. For each $j = 1,\ldots,s$, there are an index set A_j and a mapping π of A_j into the set I of projections such that $|A_j| = k$ and $\psi_j = \prod_{a\in A_j} \pi(a)$. We may take the index sets A_j to be pairwise disjoint. Let $A = \bigcup_{j=1}^s A_j$ so that $\psi = \prod_{a\in A}\pi(a)$. Now consider any $(\psi_1',\ldots,\psi_s')$ in the set S_ψ of s-tuples of elements from I^k such that $\psi = \prod_{j=1}^s \psi_j'$. Then clearly we have $\prod_{j=1}^s \psi_j' = \prod_{a\in A}\pi(a)$. Since I is independent, there is a permutation ρ of A so that $\psi_j' = \prod_{a\in A_j}\pi(\rho(a))$ for each j. Thus

$$\rho \to \Big(\prod_{a\in A_1} \pi(\rho(a)),\ldots,\prod_{a\in A_s} \pi(\rho(a))\Big)$$

maps the group P of permutations of A onto S_ψ. This map is constant on cosets of the subgroup P_0 of permutations ρ of A that map each A_j into itself. Hence

$$|S_\psi| \leqq |P/P_0| = \frac{|P|}{|P_0|} = \frac{(ks)!}{k!^s} .$$

Since $(ks)! \leqq s^s(2s)^s\cdots(ks)^s = (s^k k!)^s$, we have $|S_\psi| \leqq s^{ks}$ for all ψ and so (4) holds. □

*5.11 LEMMA. <u>Let</u> $\Omega = \mathbb{T}^I$ <u>be as in 5.6. For each</u> $k \in \mathbb{Z}^+$ <u>there is a constant</u> $A(k)$, <u>depending only on</u> k, <u>with the following property. To each</u> $\omega \in \Omega$ <u>there corresponds</u> μ_ω <u>in</u> $M(G)$ <u>such that</u>

$$\|\mu_\omega\| \leqq A(k) \tag{1}$$

<u>and</u>

$$\hat{\mu}_\omega\Big(\prod_{i\in J} \pi_i^{\varepsilon_i}\Big) = \prod_{i\in J} \omega_i \quad \underline{\text{whenever}} \quad |J| = k \quad \underline{\text{and}} \quad \varepsilon_i \in \{-1,1\}, \tag{2}$$

<u>where</u> J <u>is a subset of the index set</u> I.

[It is easy to arrange for (2) with all $\varepsilon_i = 1$ by using Riesz products. The problem is to obtain (2) in general.]

<u>Proof</u>. First we prove that given distinct real numbers $\alpha_0,\alpha_1,\ldots,\alpha_k$, there exist complex numbers $c_0,c_1,\ldots,c_k$ so that

$$\prod_{j=1}^{k}(\cos\beta_j + i\sin\beta_j) = \sum_{\ell=0}^{k} c_\ell \prod_{j=1}^{k}(\cos\beta_j + \alpha_\ell\sin\beta_j) \tag{3}$$

for all real numbers $\beta_1,\beta_2,\ldots,\beta_k$. Consider the linear system

$$\sum_{\ell=0}^{k} (\alpha_\ell)^m c_\ell = i^m \qquad (m = 0,1,\ldots,k) \tag{4}$$

of $k+1$ equations in the $k+1$ unknowns $c_0,c_1,\ldots,c_k$. The matrix of the coefficients of this system is the Vandermonde matrix in $\alpha_0,\alpha_1,\ldots,\alpha_k$, and it has nonzero determinant, namely the product of all terms $\alpha_i - \alpha_j$ with $i > j$. Hence the system (4) has a unique solution which we again denote by $c_0, c_1, \ldots, c_k$. Now let $\beta_1, \ldots, \beta_k$ be real numbers. Then we have

$$\prod_{j=1}^{k}(\cos\beta_j + i\sin\beta_j) = \sum_{m=0}^{k} i^m q_m(\beta_1, \ldots, \beta_k) \tag{5}$$

where $q_m(\beta_1, \ldots, \beta_k)$ denotes the sum of all terms in the expansion of $\prod_{j=1}^{k}(\cos\beta_j + \sin\beta_j)$ containing precisely m sine factors and $k-m$ cosine factors. From (5) and (4) we obtain

$$\begin{aligned}\prod_{j=1}^{k}(\cos\beta_j + i\sin\beta_j) &= \sum_{m=0}^{k}\sum_{\ell=0}^{k}(\alpha_\ell)^m c_\ell q_m(\beta_1, \ldots, \beta_k)\\ &= \sum_{\ell=0}^{k} c_\ell \sum_{m=0}^{k}(\alpha_\ell)^m q_m(\beta_1, \ldots, \beta_k)\\ &= \sum_{\ell=0}^{k} c_\ell \prod_{j=1}^{k}(\cos\beta_j + \alpha_\ell\sin\beta_j).\end{aligned}$$

This is equation (3). Since $\alpha_0, \alpha_1, \ldots, \alpha_k$ can be any $k+1$ distinct real numbers, we can arrange for all α_ℓ to lie in $[-1,1]$.

Now consider $\omega \in \Omega$. For each $i \in I$, we write

$$\omega_i = \cos\beta_i + i\sin\beta_i \quad\text{and}\quad a_i^{(\ell)} = \tfrac{1}{4}(\cos\beta_i + \alpha_\ell\sin\beta_i).$$

Since each $a_i^{(\ell)}$ is real and $|a_i^{(\ell)}| \leqq \frac{1}{2}$, Theorem 2.7 provides us with a (nonnegative) measure $\mu_\omega^{(\ell)}$ such that $\|\mu_\omega^{(\ell)}\| = 1$ and

$$(\mu_\omega^{(\ell)})^\wedge\Big(\prod_{i\in J}\pi_i^{\varepsilon_i}\Big) = \prod_{i\in J} a_i^{(\ell)} \tag{6}$$

for finite $J \subset I$ and $\varepsilon_i \in \{-1,1\}$. Let $\mu_\omega = \sum_{\ell=0}^{k} 4^k c_\ell \mu_\omega^{(\ell)}$ and note that $\|\mu_\omega\| \leqq 4^k\sum_{\ell=0}^{k}|c_\ell| \equiv A(k)$. If $|J| = k$, then

(6) and (3) yield

$$\hat{\mu}_\omega\Big(\prod_{i\in J} \pi_i^{\varepsilon_i}\Big) = \sum_{\ell=0}^{k} 4^k c_\ell \prod_{i\in J} a_i^{(\ell)}$$

$$= \sum_{\ell=0}^{k} c_\ell \prod_{i\in J}(\cos\beta_i + \alpha_\ell \sin\beta_i)$$

$$= \prod_{i\in J}(\cos\beta_i + i\sin\beta_i) = \prod_{i\in J} \omega_i .$$

This proves (2) as desired. □

*5.12 LEMMA. _Let_ $\Omega = \mathbb{T}^I$ _be as in_ 5.6. _For_ $k \in \mathbb{Z}^+$, _let_ D_k _consist of all characters on_ Ω _of the form_ $\prod_{i\in J} \pi_i^{\varepsilon_i}$ _where_ $|J| = k$ _and_ $\varepsilon_i \in \{-1,1\}$. _Then_ D_k _is a_ $\Lambda(q)$ _set for_ $q > 1$. _Furthermore, there is a constant_ $A(k)$, _depending only on_ k, _such that_

$$\|f\|_q \leqq A(k)\, q^{k/2}\, \|f\|_2 \tag{1}$$

for $f \in \mathrm{Trig}_{D_k}(\Omega)$ _and_ $2 < q < \infty$.

Proof. Consider f in $\mathrm{Trig}_{D_k}(\Omega)$ and for each $\omega \in \Omega$, let μ_ω be as in 5.11. From 5.11(2) we see that $\hat{\mu}_\omega \hat{\mu}_{\omega^{-1}} = 1$ on D_k and hence $f = f*\mu_\omega*\mu_{\omega^{-1}}$. Now consider the function

$$F(\omega,\omega') = F_{\omega'}(\omega) = f*\mu_\omega(\omega'), \quad (\omega,\omega') \in \Omega \times \Omega.$$

We next show that

$$\text{each } F_{\omega'} \text{ belongs to } \mathrm{Trig}_{I^k}(\Omega). \tag{2}$$

For this we use the following temporary notation. If ψ belongs to D_k, so that $\psi = \prod_{i\in J} \pi_i^{\varepsilon_i}$ where $\varepsilon_i \in \{-1,1\}$ and $|J| = k$, then we denote the corresponding character $\prod_{i\in J} \pi_i$ in I^k by ψ^+. Formula 5.11(2) tells us that

$\hat{\mu}_\omega(\psi) = \psi^+(\omega)$ for $\psi \in D_k$. Hence we have

$$F_{\omega'}(\omega) = \sum_{\psi \in D_k} \hat{f}(\psi)\hat{\mu}_\omega(\psi)\psi(\omega') = \sum_{\psi \in D_k} \hat{f}(\psi)\psi(\omega')\psi^+(\omega)$$

and so (2) is valid. This also shows that F is a trigonometric polynomial on $\Omega \times \Omega$.

Since $f = f*\mu_\omega*\mu_{\omega^{-1}}$, 5.11(1) implies

$$\|f\|_q \leqq \|f*\mu_\omega\|_q \|\mu_{\omega^{-1}}\| \leqq A(k)\|f*\mu_\omega\|_q \quad \text{for} \quad \omega \in \Omega.$$

Since F is continuous on $\Omega \times \Omega$, we can write

$$\|f\|_q^q \leqq A(k)^q \int_\Omega \|f*\mu_\omega\|_q^q d\omega = A(k)^q \int_\Omega \int_\Omega |f*\mu_\omega(\omega')|^q d\omega' \, d\omega$$

$$= A(k)^q \int_\Omega \int_\Omega |F_{\omega'}(\omega)|^q d\omega \, d\omega' = A(k)^q \int_\Omega \|F_{\omega'}\|_q^q d\omega'. \qquad (3)$$

Now we apply Theorem 5.10 to $F_{\omega'}$ in $\mathrm{Trig}_{I^k}(\Omega)$ to obtain

$$\|F_{\omega'}\|_q^2 \leqq q^k \|F_{\omega'}\|_2^2 = q^k \sum_{\chi \in I^k} |\hat{F}_{\omega'}(\chi)|^2$$

$$= q^k \sum_{\chi \in I^k} \Big| \sum_{\psi^+ = \chi} \hat{f}(\psi)\psi(\omega') \Big|^2 \leqq q^k \sum_{\psi \in D_k} |\hat{f}(\psi)|^2$$

$$= q^k \|\hat{f}\|_2^2 = q^k \|f\|_2^2.$$

From this and (3), we see at once that

$$\|f\|_q^q \leqq A(k)^q \int_\Omega q^{kq/2} \|f\|_2^q \, d\omega' = A(k)^q q^{kq/2} \|f\|_2^q,$$

i.e. (1) holds. □

Now we can prove our main theorem, due to Bonami [1970; page 359]. As usual, X denotes the character group of a compact abelian group G.

*5.13 THEOREM. <u>Let</u> E <u>be a Rider set in</u> X; <u>see 2.13</u>. <u>For</u> $k \in \mathbb{Z}^+$, <u>let</u> E_k <u>consist of all characters of the form</u> $\prod_{\chi \in S} \chi$ <u>where</u> S <u>is an asymmetric subset of</u> $E \cup E^{-1}$ <u>and</u> $|S| = k$. <u>Then</u> E_k <u>is a</u> $\Lambda(q)$ <u>set for all finite</u> $q > 1$.

Moreover, for $f \in \mathrm{Trig}_{E_k}(G)$ we have

$$\|f\|_q \leq A(k)q^{k/2}\|f\|_2 \quad \text{for} \quad 2 < q < \infty, \tag{1}$$

where $A(k)$ is a constant depending only on k and the Rider constant B of E.

Proof. Let q' denote the conjugate exponent of q. Since 5.3(v) and 5.3(vi) are equivalent, we need only show that

$$\|\hat{f}|_{E_k}\|_2 \leq A(k)\, q^{k/2}\, \|f\|_{q'} \tag{2}$$

for all $f \in L^{q'}(G)$. Since $\lim_\alpha \|f*h_\alpha - f\|_{q'} = 0$ for f in $L^{q'}(G)$ and a bounded approximate unit $\{h_\alpha\}$ for $L^1(G)$ consisting of trigonometric polynomials, it suffices to establish (2) for all $f \in \mathrm{Trig}(G)$. Observe also that we may suppose without loss of generality that E is asymmetric.

Let $B \geq 1$ satisfy $R_s(E,1) \leq B^s$ for all $s \geq 0$. We fix f in $\mathrm{Trig}(G)$ and consider a finite subset F of E such that $|F| \geq k$ and $\mathrm{Supp}(\hat{f}) \cap E_k \subset F_k$. This arranges for $\mathrm{Supp}(\hat{f}) \cap E_k = \mathrm{Supp}(\hat{f}) \cap F_k$ so that

$$\|\hat{f}|_{E_k}\|_2 = \|\hat{f}|_{F_k}\|_2. \tag{3}$$

The auxiliary group that we use is $\Omega = \mathbb{T}^F$. For each $\chi \in F$, π_χ will denote the corresponding projection on Ω: $\pi_\chi(\omega) = \omega_\chi$ for $\omega \in \Omega$. For each $s \geq 0$, let D_s consist of all products $\prod_{\chi\in T} \pi_\chi \prod_{\chi\in S\setminus T} \pi_\chi^{-1}$ where $T \subset S \subset F$ and $|S| = s$; these sets are the same as the sets D_k in Lemma 5.12. For each ψ in D_s, the corresponding character $\prod_{\chi\in T} \chi \prod_{\chi\in S\setminus T} \chi^{-1}$ in X will be denoted by ψ'. Note that

every character in F_k has the form ψ' for some $\psi \in D_k$. (4)

We next consider the following Riesz product on $G \rtimes \Omega$:

$$p(x,\omega) = \prod_{\chi\in F}[1 + (2B)^{-1}\mathrm{Re}(\chi(x)\pi_\chi(\omega))].$$

We expand p using 2.4(3):

$$\begin{aligned} p(x,\omega) &= \sum_{S\subset F} \prod_{\chi\in S}[(2B)^{-1}\mathrm{Re}(\chi(x)\pi_\chi(\omega))] \\ &= \sum_{S\subset F} \prod_{\chi\in S}(4B)^{-1}[\chi(x)\pi_\chi(\omega) + \chi^{-1}(x)\pi_\chi^{-1}(\omega)] \\ &= \sum_{S\subset F}(4B)^{-|S|}\sum_{T\subset S}\prod_{\chi\in T}[\chi(x)\pi_\chi(\omega)]\prod_{\chi\in S\setminus T}[\chi^{-1}(x)\pi_\chi^{-1}(\omega)] \\ &= \sum_{s=0}^{|F|}(4B)^{-s}\sum_{\psi\in D_s}\psi'(x)\psi(\omega). \end{aligned} \tag{5}$$

For each $\omega\in\Omega$, p_ω denotes the trigonometric polynomial on G given by $p_\omega(x) = p(x,\omega)$. The definition of $R_s(E,1)$ shows that $\{\psi\in D_s : \psi' = 1\}$ has at most $2^s R_s(E,1) \leqq (2B)^s$ terms. [The 2^s term arises because characters of order 2 can be in T or $S\setminus T$ in the product $\prod_{\chi\in T}\chi\cdot\prod_{\chi\in S\setminus T}\chi^{-1}$.] Since $p_\omega \geqq 0$, we conclude from (5) that

$$\|p_\omega\|_1 = \hat{p}_\omega(1) \leqq \sum_{s=0}^{|F|}(4B)^{-s}(2B)^s < 2.$$

It follows that

$$\|p_\omega * f\|_{q'} \leqq \|p_\omega\|_1\|f\|_{q'} \leqq 2\|f\|_{q'} \quad \text{for } \omega\in\Omega. \tag{6}$$

Since $\psi' * f(x) = \psi'(x)\hat{f}(\psi')$ for $\psi'\in X$ and $x\in G$, we see from (5) that

$$p_\omega * f(x) = \sum_{s=0}^{|F|}(4B)^{-s}\sum_{\psi\in D_s}\psi'(x)\psi(\omega)\hat{f}(\psi'). \tag{7}$$

For $x\in G$, we define F_x on Ω by $F_x(\omega) = p_\omega * f(x)$. From (7) we see that

$$\hat{F}_x(\psi) = (4B)^{-k}\psi'(x)\hat{f}(\psi') \quad \text{for } \psi\in D_k. \tag{8}$$

Now by (6) we have

$$2^{q'}\|f\|_{q'}^{q'} = \int_{\Omega} 2^{q'}\|f\|_{q'}^{q'}\,d\omega \geqq \int_{\Omega}\|p_{\omega}*f\|_{q'}^{q'}\,d\omega$$
$$= \int_{\Omega}\int_{G}|p_{\omega}*f(x)|^{q'}d\lambda(x)d\omega$$
$$= \int_{G}\int_{\Omega}|F_x(\omega)|^{q'}d\omega\, d\lambda(x) = \int_{G}\|F_x\|_{q'}^{q'}d\lambda(x). \qquad (9)$$

Since 5.12(1) holds for D_k and some constant $A(k)$, 5.3(vi) shows that

$$\|\hat{F}|_{D_k}\|_2 \leqq A(k)\, q^{k/2}\, \|F\|_q,$$

for $F \in \mathrm{Trig}(\Omega)$. Applying this to F_x and using (8), we obtain

$$\sum_{\psi\in D_k}(4B)^{-2k}|\hat{f}(\psi')|^2 \leqq A(k)^2 q^k \|F_x\|_{q'}^2.$$

The observation in (4) implies that

$$\sum_{\chi\in F_k}(4B)^{-2k}|\hat{f}(\chi)|^2 \leqq A(k)^2\, q^k\, \|F_x\|_{q'}^2,$$

i.e.

$$\|\hat{f}|_{F_k}\|_2 \leqq (4B)^k A(k)\, q^{k/2}\, \|F_x\|_{q'}.$$

This inequality and (9) lead to

$$\|\hat{f}|_{F_k}\|_2 \leqq 2(4B)^k A(k)\, q^{k/2}\, \|f\|_{q'}.$$

A glance at (3) shows that (2) has been established where $A(k)$ is replaced by $2(4B)^k A(k)$. □

*5.14 COROLLARY. <u>If</u> X <u>is infinite, then</u> X <u>contains a non-Sidon set that is a</u> $\Lambda(q)$ <u>set for all finite</u> $q > 1$.

<u>Proof</u>. Let E be an infinite Rider set in X. Such sets always exist by Theorem 2.8. By Theorem 5.13, E_2 is a $\Lambda(q)$ set for all finite $q > 1$. If D_1 and D_2 are infinite disjoint subsets of E and $D_1 \cup D_2$ is asymmetric, then

$D_1D_2 \subset E_2$ and so E_2 cannot be a Sidon set by Theorem 1.4. □

*5.15 COROLLARY. If D_1 and D_2 are disjoint and $D_1 \cup D_2$ is an asymmetric Rider set, then D_1D_2 is a $\Lambda(q)$ set for all finite $q > 1$.

*5.16 REMARKS. (a) Corollary 5.15 and Theorem 5.13 are no longer valid if "Rider set" is replaced by "the union of two Rider sets". This can be seen by examining the following subset of $\mathbb{Z}$: $E = D_1 \cup D_2$ where $D_1 = \{10^n + n : n \in \mathbb{Z}^+\}$ and $D_2 = \{-10^n : n \in \mathbb{Z}^+\}$. Then $D_1 + D_2 \supset \mathbb{Z}^+$ and so neither this set nor E_2 is a $\Lambda(q)$ set for any $q > 1$. The sets D_1 and D_2 are Rider sets, since D_1 and $-D_2$ are in fact Hadamard sets $\{n_k\}_{k=1}^{\infty}$ with $n_{k+1}/n_k \geqq 3$ for all k; see 2.11. This example is due to Sam Ebenstein.

(b) Meyer [1968; page 558] proved that if E and F are Hadamard sets in $\mathbb{Z}$, then $E + F$ is a $\Lambda(q)$ set for all finite $q > 1$. He also proved that if $E = \{n_k\}_{k=1}^{\infty}$ is Hadamard lacunary in $\mathbb{Z}$ with $n_{k+1}/n_k \geqq 3$ for all k, then E_k (as defined in 5.13) is a $\Lambda(q)$ set for all finite $q > 1$. Bonami [1968a; page 196] proved that if E is an independent set in the dual of $\mathbb{Z}(2)^{\aleph_0}$, then E^k is a $\Lambda(q)$ set for all $k \in \mathbb{Z}^+$ and all finite $q > 1$. This result is closely related to Theorem 5.10. See also Blei [1974b].

(c) Corollary 5.14 was proved for $X = \mathbb{Z}$ by Rudin [1960; page 222] by different means. Corollary 5.14 was proved as stated by Edwards, Hewitt and Ross [1972a] via technical constructions which were complicated elaborations of Rudin's original construction. The present proof is more perspicuous and gives a better result: the sets constructed by Edwards, Hewitt and Ross were just barely big enough to violate Sidonicity, while Bonami's $\Lambda(q)$ sets E_k are rather large. A proof of 5.14 similar to our proof is sketched by Ebenstein [1972].

(d) If E is an infinite $\Lambda(q)$ set in X for some $q > 1$, then its characteristic function ξ_E cannot have the form $\hat{\mu}$ for some (idempotent) measure $\mu \in M(G)$. This follows from 5.3(ii) and the fact that $\hat{f} \in c_0(X)$ for $f \in L^q(G) \subset L^1(G)$.

If E is a dissociate set, a simple direct proof of this can be given; this was observed by O. C. McGehee. Assume that $\hat{\mu} = \xi_E$ where $\mu \in M(G)$ and E is an infinite dissociate set. Select distinct $\chi_1, \chi_2, \ldots$ from E and let p_k be the Riesz product $\prod_{j=1}^k g_j$ where $g_j = 1 + \frac{i}{2\sqrt{k}}(\chi_j + \chi_j^{-1})$ if $\chi_j^2 \neq 1$ and $g_j = 1 + \frac{i}{2\sqrt{k}}\chi_j$ if $\chi_j^2 = 1$. Then we have

$$\int_G p_k \, d\mu = \sum_{j=1}^k \hat{p}_k(\chi_j) = \sqrt{k}\, i/2. \tag{1}$$

We also have

$$\|p_k\|_u \leqq \prod_{j=1}^k \left|1 + \frac{i}{\sqrt{k}}\right| = \prod_{j=1}^k \left(1 + \frac{1}{k}\right)^{\frac{1}{2}}$$

$$\leqq \prod_{j=1}^k \left(1 + \frac{1}{k}\right) = \left(1 + \frac{1}{k}\right)^k \to e \quad \text{as} \quad k \to \infty,$$

which is incompatible with (1).

Chapter 6

ARITHMETIC PROPERTIES*

There are a number of theorems that show Sidon sets must satisfy certain arithmetic conditions. We will prove a few of these in this chapter. One of the most useful properties of Sidon sets was established in Theorem 1.4. We will give a second proof of this theorem in Corollary 6.6 based on one of the sharpest and best known results of this kind; see 6.5. Corollary 6.5 asserts that Sidon sets cannot contain large portions of certain sets that are generalized arithmetic progressions. We define these sets in 6.2, but first we prove a useful lemma.

*6.1 LEMMA. <u>For</u> χ <u>in</u> X <u>and an integer</u> $N > 0$, <u>we define</u>

$$g_\chi(x) = \sum_{n\in\mathbb{Z}} \exp\{-(\tfrac{n}{N})^2\}\chi^n(x) \quad \underline{\text{for}} \quad x \in G.$$

<u>Then</u> g_χ <u>is a nonnegative function on</u> G <u>and</u>

$$\|g_\chi\|_u \leqq 1 + \sqrt{\pi}N. \tag{1}$$

<u>Proof</u>. We use the classical fact that the function $u \to \exp\{-\frac{1}{2}u^2\}$ on $\mathbb{R}$ is its own Fourier transform:

$$(2\pi)^{-\frac{1}{2}}\int_{-\infty}^{\infty} \exp\{-\tfrac{1}{2}u^2 - iuv\}du = \exp\{-\tfrac{1}{2}v^2\} \quad \text{for} \quad v \in \mathbb{R}. \tag{2}$$

A straightforward proof of this is indicated in [Kz; page 127 and page 129, Exercise 4]. Proofs involving contour integrals are given in [HR; Vol. II, 31.9] and [Z; Vol. II, page 252.]

---

* None of the results given in this chapter will be referred to in the main development of the notes.

In particular, $v \to \exp\{-(\frac{v}{N})^2\}$ is the Fourier transform of $u \to 2^{-\frac{1}{2}}N\exp\{-(\frac{uN}{2})^2\}$. Since the second function is nonnegative on $\mathbb{R}$, the easy half of Bochner's theorem tells us that the first function is positive-definite. It is clear from the definition of positive-definite function that its restriction $n \to \exp\{-(\frac{n}{N})^2\}$ to $\mathbb{Z}$ is also positive-definite. Hence, by Bochner's theorem again, the function

$$h(t) = \sum_{n\in\mathbb{Z}} \exp\{-(\tfrac{n}{N})^2\}\exp\{int\}$$

is nonnegative on $[0,2\pi)$.

Now consider χ in X and x in G. For some t in $[0,2\pi)$ we have $\chi(x) = e^{it}$ and so $g_\chi(x) = h(t) \geqq 0$. This holds for all $x \in G$ and so g_χ is a nonnegative function on G.

To check (1), we observe that

$$\|g_\chi\|_u \leqq \sum_{n\in\mathbb{Z}} \exp\{-(\tfrac{n}{N})^2\} = 1 + 2\sum_{n=1}^{\infty}\exp\{-(\tfrac{n}{N})^2\}$$

$$\leqq 1 + 2\sum_{n=1}^{\infty}\int_{n-1}^{n} \exp\{-(\tfrac{x}{N})^2\}dx$$

$$= 1 + 2\int_0^{\infty} \exp\{-(\tfrac{x}{N})^2\}dx = 1 + \sqrt{\pi}\,N.$$

For the last equality, put $v = 0$ in (2). □

*6.2 DEFINITION. Consider positive integers N and s and consider $(\chi_1,\chi_2,\ldots,\chi_s)$ in X^s. For $1 \leqq r < \infty$, we define

$$A_r(N,\chi_1,\chi_2,\ldots,\chi_s) = \{\prod_{j=1}^{s} \chi_j^{n_j} : \sum_{j=1}^{s} |n_j|^r \leqq N^r\}. \qquad (1)$$

This may be viewed as a generalized s-dimensional arithmetic progression. Only the values $r = 1$ and $r = 2$ will interest us.

Our main results are given in the next theorem and its corollaries. The history of these results is discussed in 6.7.

*6.3 THEOREM. Let $E \subset X$ be a $\Lambda(q)$ set for some $q > 2$, and let η_q be a constant so that $\|f\|_q \leqq \eta_q \|f\|_2$ for all f in $\mathrm{Trig}_E(G)$; see 5.3(1). Then we have

$$|A_2 \cap E| \leqq e^2 \eta_q^2 (1 + \sqrt{\pi}\, N)^{2s/q} \tag{1}$$

for all sets $A_2 = A_2(N, \chi_1, \chi_2, \ldots, \chi_s)$.

Proof. Let N and $\chi_1, \chi_2, \ldots, \chi_s$ be fixed. We will use the auxiliary function $g = \prod_{j=1}^{s} g_{\chi_j}$ where g_{χ_j} is as defined in Lemma 6.1. Applying Fubini's theorem on $\mathbb{Z}^s$, we have

$$g(x) = \prod_{j=1}^{s}\Big[\sum_{n\in\mathbb{Z}} \exp\{-(\tfrac{n}{N})^2\}\, \chi_j^n(x)\Big]$$

$$= \sum_{(n_1,\ldots,n_s)\in\mathbb{Z}^s} \Big[\prod_{j=1}^{s} \exp\{-(\tfrac{n_j}{N})^2\}\, \chi_j^{n_j}(x)\Big]. \tag{2}$$

For each $\vec{n} = (n_1, \ldots, n_s)$ in $\mathbb{Z}^s$, we define

$$\phi(\vec{n}) = \prod_{j=1}^{s} \chi_j^{n_j}.$$

Then clearly ϕ is a homomorphism of $\mathbb{Z}^s$ into X. Moreover, (2) can be rewritten as

$$g = \sum_{\vec{n}\in\mathbb{Z}^s} \Big[\exp\{-\sum_{j=1}^{s} (\tfrac{n_j}{N})^2\} \prod_{j=1}^{s} \chi_j^{n_j}\Big]$$

$$= \sum_{\vec{n}\in\mathbb{Z}^s} \Big[\exp\{-\Big\|\tfrac{\vec{n}}{N}\Big\|^2\}\, \phi(\vec{n})\Big] \tag{3}$$

[Norms in this proof without subscripts always signify 2-norms in $\mathbb{R}^s$.] From (3) we can read off $\hat{g}$:

$$\hat{g}(\chi) = \begin{cases} \displaystyle\sum_{\vec{n}\in\phi^{-1}(\chi)} \exp\{-\Big\|\tfrac{\vec{n}}{N}\Big\|^2\} & \text{for } \chi \in \phi(\mathbb{Z}^s) \\ 0 \quad \text{for } \chi \notin \phi(\mathbb{Z}^s). \end{cases}$$

Next we prove that

$$\hat{g}(1) \leqq e\,\hat{g}(\chi) \quad \text{for all} \quad \chi \in A_2. \tag{4}$$

Let H denote the kernel of ϕ, and let H^+ denote an asymmetric subset of $H \setminus \{0\}$ so that $H = \{0\} \cup H^+ \cup (-H^+)$ and the union is disjoint. Now consider any χ in A_2. Then $\chi = \phi(\vec{n})$ for some $\vec{n} \in \mathbb{Z}^s$ such that $\|\vec{n}\| \leqq N$. Since $\phi^{-1}(\chi) = \vec{n} + H$, we have

$$\hat{g}(\chi) = \sum_{\vec{m} \in H} \exp\{-\left\| \frac{\vec{n} + \vec{m}}{N} \right\|^2\}$$

$$= \exp\{-\left\| \frac{\vec{n}}{N} \right\|^2\} + \sum_{\vec{m} \in H^+} [\exp\{-\left\| \frac{\vec{n} + \vec{m}}{N} \right\|^2\} + \exp\{-\left\| \frac{\vec{n} - \vec{m}}{N} \right\|^2\}].$$

Using the identity

$$\exp\{-\left\| \frac{\vec{n} \pm \vec{m}}{N} \right\|^2\} = \exp\{-\left\| \frac{\vec{n}}{N} \right\|^2 - \left\| \frac{\vec{m}}{N} \right\|^2\} \exp\{\mp 2 \langle \vec{n}, \vec{m} \rangle N^{-2}\},$$

we obtain

$$\hat{g}(\chi) = \exp\{-\left\| \frac{\vec{n}}{N} \right\|^2\} [1 + 2 \sum_{\vec{m} \in H^+} \exp\{-\left\| \frac{\vec{m}}{N} \right\|^2\} \cosh(2 \langle \vec{n}, \vec{m} \rangle N^{-2})].$$

Let $\chi = 1$ and $\vec{n} = \vec{0}$; then

$$\hat{g}(1) = 1 + 2 \sum_{\vec{m} \in H^+} \exp\{-\left\| \frac{\vec{m}}{N} \right\|^2\}$$

and so

$$\hat{g}(\chi) \geqq \exp\{-\left\| \frac{\vec{n}}{N} \right\|^2\} \hat{g}(1).$$

Since $\|\vec{n}\| \leqq N$, we see that (4) holds.

Next we estimate $\|g\|_{q'}$ where q' is the conjugate exponent of q. By Hölder's inequality 5.1, we have

$$\|g\|_{q'} \leqq \|g\|_1^{1/q'} \|g\|_\infty^{1/q}.$$

In view of Lemma 6.1, g is a nonnegative function and so $\|g\|_1 = \hat{g}(1) \geqq 1$. And from inequality 6.1(1), we obtain $\|g\|_\infty \leqq (1 + \sqrt{\pi} N)^s$. These observations show that

$$\|g\|_{q'} \leqq \hat{g}(1)^{1/q'} (1 + \sqrt{\pi} N)^{s/q} \leqq \hat{g}(1)(1 + \sqrt{\pi} N)^{s/q}. \tag{5}$$

To complete the proof, let f be the trigonometric polynomial whose Fourier transform $\hat{f}$ is the characteristic function of $A_2 \cap E$, and let $\alpha = |A_2 \cap E|$. Then (4), Parseval's relation, the standard Hölder inequality, and (5) in turn lead to

$$\alpha\hat{g}(1) = \sum_{\chi \in X} \hat{f}(\chi)\hat{g}(1) \leqq e \sum_{\chi \in X} \hat{f}(\chi)\hat{g}(\chi) = e \int_G fg\, d\lambda$$

$$\leqq e\|f\|_q \|g\|_{q'} \leqq e\|f\|_q \hat{g}(1)(1 + \sqrt{\pi}N)^{s/q}. \qquad (6)$$

Since $\|f\|_q \leqq \eta_q \|f\|_2 = \eta_q \|\hat{f}\|_2 = \eta_q \alpha^{\frac{1}{2}}$, (6) shows that

$$\alpha \leqq e\, \eta_q\, \alpha^{\frac{1}{2}}(1 + \sqrt{\pi}N)^{s/q},$$

i.e. $|A_2 \cap E| = \alpha \leqq e^2 \eta_q^2 (1 + \sqrt{\pi}N)^{2s/q}$. □

*6.4 COROLLARY. *If* $E \subset X$ *is a* $\Lambda(q)$ *set for some* $q > 2$, *then inequality 6.3(1) also holds for all sets* $A_1(N, \chi_1, \chi_2, \ldots, \chi_s)$.

Proof. $A_1(N, \chi_1, \chi_2, \ldots, \chi_s) \subset A_2(N, \chi_1, \chi_2, \ldots, \chi_s)$. □

We now give the corresponding results for Sidon sets.

*6.5 COROLLARY. *If* $E \subset X$ *is a Sidon set with Sidon constant* κ, *then*

$$|A \cap E| \leqq 2\kappa^2 e^3 s \log(1 + \sqrt{\pi}N) \leqq K s \log N \qquad (1)$$

for sets $A = A_r(N, \chi_1, \chi_2, \ldots, \chi_s)$, $r = 1,2$, *where* K *is a constant depending only on* κ.

Proof. By Theorem 5.8, E is a $\Lambda(q)$ set for $q > 1$ and $\|f\|_q \leqq \kappa\sqrt{q}\, \|f\|_2$ for $q > 2$ and $f \in \mathrm{Trig}_E(G)$. Thus 6.3(1) holds with $\eta_q = \kappa\sqrt{q}$:

$$|A \cap E| \leqq e^2 \kappa^2 q(1 + \sqrt{\pi}N)^{2s/q}$$

$$= e^2 \kappa^2 q \exp\{(2s/q)\log(1 + \sqrt{\pi}N)\}.$$

For $q = 2s\log(1+\sqrt{\pi}N)$ this inequality becomes inequality (1); note that $q \geqq 2\log(1+\sqrt{\pi}) \doteq 2.039 > 2$. □

Next we restate and reprove Theorem 1.4.

*6.6 COROLLARY. <u>If</u> E <u>is a Sidon set in</u> X, <u>then</u>

$$\sup\{\min(|A|,|B|) : AB \subset E\} < \infty. \qquad (1)$$

<u>Proof</u>. As in 1.4 we assume (1) fails, fix $n > 1$, and obtain disjoint sets $D_1 \subset A$ and $D_2 \subset B$ so that $|D_1| = |D_2| = n$, $D = D_1D_2 \subset E$ and $|D| = n^2$. Let $s = 2n$ and write $D_1 \cup D_2$ as $\chi_1, \chi_2, \ldots, \chi_s$. Then clearly we have $A_1 = A_1(2, \chi_1, \chi_2, \ldots, \chi_s) \supset D$ and so by Corollary 6.5

$$n^2 = |D| \leqq |A_1 \cap E| \leqq Ks\log 2 = 2Kn\log 2.$$

As n is arbitrary, a contradiction ensues. □

Before giving some applications of Corollary 6.5 we pause to give a brief history of 6.3 - 6.6.

*6.7 REMARKS. The basic result in Corollary 6.5 is due to Kahane [1957; page 312] for $r = 1$. It is also proved in [Ka; page 58]. Both proofs are given for $X = \mathbb{Z}$ and involve bounds for random trigonometric polynomials. The present treatment of 6.1 - 6.5 is due to Benke [1971] and [1972]. In particular, Theorem 6.3 for A_2 sets and Corollary 6.5 for $r = 2$ are due to Benke. Benke calls a set $E \subset X$ a Λ <u>set</u> provided a constant κ exists such that $\|f\|_q \leqq \kappa\sqrt{q}\,\|f\|_2$ for all f in $\mathrm{Trig}_E(G)$ and $2 < q < \infty$. It is evident from the proof of 6.5 that 6.5(1) holds for Λ sets. This result is also due to Benke, since Kahane's proof relies on 1.3(vii). This improvement might be vacuous, however, since no non-Sidon Λ sets are known. Benke [1971],[1974c] gives a detailed study of Λ sets.

The proof of Corollary 6.6 from Corollary 6.5 is advised in Exercise 4, page 61, of [Ka] and executed in Benke [1971]. Yet another proof of 6.6 is given by Edwards, Hewitt and Ross [1972a; 3.5]; it was communicated to them by A. Figa-Talamanca.

*6.8 EXAMPLE. We now specialize the results in 6.3 - 6.5 to arithmetic progressions in $\mathbb{Z}$ and obtain results similar to those that appear in Rudin [1960]; see 6.12(a). An arithmetic progression in $\mathbb{Z}$ is a set of the form $\{a+b, a+2b, \ldots, a+Nb\}$ where $a \in \mathbb{Z}$ and $b, N \in \mathbb{Z}^+$. For $E \subset \mathbb{Z}$ and positive integers N, we let $\alpha_E(N)$ be the largest integer α such that some arithmetic progression of N terms contains α terms of E.

An arithmetic progression in $\mathbb{Z}$ with $2N+1$ terms can be written as $a + A_1(N,b)$ for some $a, b \in \mathbb{Z}$; see 6.2(1). Let E be a $\Lambda(q)$ set in $\mathbb{Z}$ for some $q > 2$. Then $E - a$ has the same property and so Corollary 6.4 tells us that

$$|[a + A_1(N,b)] \cap E| = |A_1(N,b) \cap (E-a)| \leqq e^2 \eta_q^2 (1 + \sqrt{\pi} N)^{2/q}. \tag{1}$$

This implies that

$$\alpha_E(2N+1) \leqq e^2 \eta_q^2 (1 + \sqrt{\pi} N)^{2/q}; \tag{2}$$

compare 6.12(3). Inequality 6.5(1) tells us that if E is a Sidon set in $\mathbb{Z}$ with Sidon constant κ, then

$$\alpha_E(2N+1) \leqq 2\kappa^2 e^3 \log(1 + \sqrt{\pi} N) \leqq K \log N \tag{3}$$

for some constant K; compare 6.12(4).

We emphasize that a consequence of (2) is this: If $E \subset \mathbb{Z}$ is a $\Lambda(q)$ set for some $q > 2$, then E cannot contain arbitrarily long arithmetic progressions. Of course this comment applies also to Sidon sets. Further remarks concerning arithmetic progressions and lacunary sets appear in 6.12.

As we have seen, generalized s-dimensional arithmetic progressions provide a useful generalization of arithmetic progressions in $\mathbb{Z}$. A little different approach was taken by Edwards, Hewitt and Ross [1972a], who observed that arithmetic progressions A in $\mathbb{Z}$ also have the property that $|A+A-A|$ is small compared to $|A|^3$. This motivated the next definition.

*6.9 DEFINITION. Consider a real number $M \geq 1$. A finite set $A \subset X$ is said to be a <u>test set of order</u> M provided

$$|A^2A^{-1}| \leq M|A|. \tag{1}$$

It is easy to verify that arithmetic progressions in $\mathbb{Z}$ are test sets of order 3; see 6.12(a). Finite subgroups of any group X are obviously test sets of order 1.

*6.10 THEOREM. <u>Let</u> $E \subset X$ <u>be a</u> $\Lambda(q)$ <u>set for some</u> $q > 2$, <u>and let</u> η_q <u>be a constant such that</u> $\|f\|_q \leq \eta_q \|f\|_2$ <u>for all</u> $f \in \mathrm{Trig}_E(G)$. <u>Then we have</u>

$$|A \cap E| \leq \eta_q^2 M\, |A|^{2/q} \tag{1}$$

<u>for all test sets</u> A <u>of order</u> M.

<u>Proof</u>. Let g be the trigonometric polynomial on G such that $\hat{g} = |A|^{-1}\xi_A * \xi_{AA^{-1}}$, where ξ_S denotes the characteristic function of S. For $\chi \in X$, we have

$$\hat{g}(\chi) = |A|^{-1} \sum_{\psi \in X} \xi_A(\chi\psi^{-1})\xi_{AA^{-1}}(\psi)$$

$$= |A|^{-1} \sum_{\psi \in X} \xi_{\chi A^{-1}}(\psi)\xi_{AA^{-1}}(\psi) = |A|^{-1}|\chi A^{-1} \cap AA^{-1}|.$$

Hence we have

$$\hat{g}(\chi) = 1 \quad \text{for} \quad \chi \in A, \tag{1}$$

$$\hat{g}(\chi) = 0 \quad \text{for} \quad \chi \notin A^2A^{-1}, \tag{2}$$

$$0 \leqq \hat{g}(\chi) \leqq 1 \quad \text{for all} \quad \chi \in X. \tag{3}$$

Next we estimate $\|g\|_{q'}$ where q' is the conjugate exponent of q. Since

$$\|g\|_1 = \| \, |A|^{-1}(\xi_A)\hat{}(\xi_{AA^{-1}})\hat{}\,\|_1 \leqq |A|^{-1}\|(\xi_A)\hat{}\|_2\|(\xi_{AA^{-1}})\hat{}\|_2$$

$$= |A|^{-1}\|\xi_A\|_2\|\xi_{AA^{-1}}\|_2 = |A|^{-\frac{1}{2}}|AA^{-1}|^{\frac{1}{2}},$$

6.9(1) implies that

$$\|g\|_1 \leqq M^{\frac{1}{2}}. \tag{4}$$

Also from (2) and (3) we see that

$$\|g\|_2^2 = \|\hat{g}\|_2^2 \leqq |A^2A^{-1}| \leqq M|A|$$

and so

$$\|g\|_2 \leqq M^{\frac{1}{2}}|A|^{\frac{1}{2}}. \tag{5}$$

Hölder's inequality 5.1, (4) and (5) now give

$$\|g\|_{q'} \leqq \|g\|_1^{1-\frac{2}{q}} \|g\|_2^{\frac{2}{q}} \leqq M^{\frac{1}{2}}|A|^{1/q}. \tag{6}$$

Next let $f = \sum_{\chi \in A \cap E} \chi \in \mathrm{Trig}_E(G)$. Then, just as in 6.3(6), we have

$$|A \cap E| = \sum_{\chi \in X} \hat{f}(\chi)\hat{g}(\chi) \leqq \|f\|_q \|g\|_{q'} \leqq \eta_q \|f\|_2 \|g\|_{q'}$$

$$= \eta_q \|\hat{f}\|_2 \|g\|_{q'} = \eta_q |A \cap E|^{\frac{1}{2}} \|g\|_{q'} .$$

This inequality and (6) now imply that

$$|A \cap E| \leqq \eta_q^2 \|g\|_{q'}^2 \leqq \eta_q^2 M \, |A|^{2/q}. \quad \square$$

*6.11 COROLLARY. <u>If</u> $E \subset X$ <u>is a Sidon set with Sidon constant</u> $\kappa \geqq 1$, <u>then</u>

$$|A \cap E| \leqq 2\kappa^2 e M \log|A| \tag{1}$$

<u>for test sets</u> A <u>of order</u> M <u>such that</u> $|A| \geqq 2$.

<u>Proof</u>. If $|A| = 2$, simply observe that $2 \leqq 2e\log 2$. If $|A| \geqq 3$, let $q = 2\log|A| > 2$ and recall that $\|f\|_q \leqq \kappa\sqrt{q}\,\|f\|_2$ for $f \in \mathrm{Trig}_E(G)$ by Theorem 5.8. Inequality (1) follows directly from 6.10(1) with $q = 2\log|A|$ and $\eta_q = \kappa\sqrt{q}$. □

Theorem 6.10 and Corollary 6.11 are taken from Edwards, Hewitt and Ross [1972a], but many of their special cases had been known before.

*6.12 EXAMPLES. (a) For $E \subset \mathbb{Z}$ and $N \geqq 1$, let $\alpha_E(N)$ be as defined in 6.8. We contend that arithmetic progressions A in $\mathbb{Z}$ are test sets of order 3. In fact, if $A = \{a + b, a + 2b, \ldots, a + Nb\}$, then

$$A + A - A = \{a + kb : -N + 2 \leqq k \leqq 2N - 1\}$$

and so

$$|A + A - A| = 3N - 2 < 3N = 3|A|.$$

If $E \subset \mathbb{Z}$ is a $\Lambda(q)$ set with $q > 2$, then by 6.10(1),

$$|A \cap E| \leqq 3\eta_q^2\,|A|^{2/q} \tag{1}$$

for arithmetic progressions A in $\mathbb{Z}$. If E is a Sidon set, then 6.11(1) shows that

$$|A \cap E| \leqq 6\kappa^2 e\log|A| \tag{2}$$

for arithmetic progressions A in $\mathbb{Z}$. In terms of $\alpha_E(N)$ defined in 6.8, (1) and (2) assert that

$$\alpha_E(N) \leqq 3\eta_q^2\,N^{2/q} \tag{3}$$

and

$$\alpha_E(N) \leqq 6\kappa^2 e\log N, \tag{4}$$

respectively. Inequalities (3) and (4) are tiny improvements on (3.5.2) and (3.6.2) in Rudin [1960].

The set $E = \{2^k : k = 0,1,\ldots\}$ is a Sidon set for which

$$\alpha_E(2^k) \geqq k+1 > \frac{\log 2^k}{\log 2}$$

and so $\log N$ in (4) cannot be replaced by any function of N which is $o(\log N)$.

Let $A(s) = \{1,2,3,\ldots,2^s\}$, $s \geqq 1$. If E is a Sidon set in $\mathbb{Z}$, then (2) gives

$$|A(s) \cap E| \leqq (6\varkappa^2 e \log 2)\cdot s.$$

This is sharper than the corresponding inequality in [LP; page 58].

(b) If $E \subset \mathbb{Z}$ is a Sidon set and $0 \notin E$, then $\sum_{x \in E} \frac{1}{|x|} < \infty$. Exercise.

(c) Let E be a Sidon set in any group X with Sidon constant $\varkappa$. Then 6.11(1) implies that

$$|A \cap E| \leqq 2\varkappa^2 e \log|A| \tag{5}$$

for all nontrivial finite subgroups A of X. For $G = \mathbb{Z}(p)^I$ where p is prime, this result is proved by Bonami [1968a]; see also [LP; page 57]. In this case, subgroups of X are vector spaces over $\mathbb{Z}(p)$. If A is a finite subgroup of X, then $|A| = p^m$ where m is the vector space dimension of A. Hence by (5),

$$|A \cap E| \leqq (2\varkappa^2 e \log p)\dim(A) \leqq K \dim(A) \tag{6}$$

for finite vector subspaces A of X. A similar result is given in Benke [1972].

The case of $G = \mathbb{Z}(p)^I$ is of considerable interest. Malliavin-Brameret and Malliavin [1967] showed that Sidon sets in X are finite unions of independent sets. This result depends on a vector space theorem due to Horn [1955]; see also Rado [1962]. These results are presented in [LP; pp. 73-75].

For the remainder of this chapter, we assume that X is countable and that $\{X_n\}_{n=1}^{\infty}$ is an increasing sequence of finite subsets whose union is X. As in 1.5, $U(G)$ will denote the family of functions f in $C(G)$ satisfying $\lim_{n\to\infty}\|s_nf - f\|_u = 0$ where $s_nf = \Sigma_{\chi\in X_n} \hat{f}(\chi)\chi$. A set $E \subset X$ will be called a <u>set of uniform convergence</u> or a UC-<u>set</u> if $C_E(G) \subset U(G)$. Thus Sidon sets are UC-sets. Figà-Talamanca [1970] first exhibited UC-sets that are not Sidon sets. Pedemonte [1974] contains a detailed study of UC-sets; see 6.15. The treatment that follows is based on these two papers.

*6.13 LEMMA. <u>A set</u> $E \subset X$ <u>is a UC-set if and only if there are a constant</u> $K > 0$ <u>and a sequence</u> $\{\mu_n\}_{n=1}^{\infty}$ <u>in</u> $M(G)$ <u>such that</u>

$$\|\mu_n\| \leqq K \quad \underline{\text{for all}} \quad n, \tag{1}$$

$$\hat{\mu}_n(\psi) = 1 \quad \underline{\text{for}} \quad \psi \in E \cap X_n, \tag{2}$$

$$\hat{\mu}_n(\psi) = 0 \quad \underline{\text{for}} \quad \psi \in E \setminus X_n. \tag{3}$$

<u>Proof</u>. Suppose that E is a UC-set. For $n = 1,2,\ldots,$ let $T_n(f) = \Sigma_{\chi\in X_n} \hat{f}(\chi)$ for $f \in C_E(G)$. Each T_n is a bounded linear functional on $C_E(G)$ and for each f in $C_E(G)$ we have

$$\begin{aligned} |T_n(f)| &= |s_nf(0)| \leqq |s_nf(0) - f(0)| + |f(0)| \\ &\leqq \|s_nf - f\|_u + |f(0)| \end{aligned}$$

and so $\sup_n |T_n(f)| < \infty$. By the uniform boundedness principle, there is a constant $K > 0$ such that $\|T_n\| \leq K$ for all n. By the Hahn-Banach theorem, each T_n can be extended to a linear functional on $C(G)$, also denoted by T_n, where $\|T_n\| \leq K$. By the Riesz representation theorem, there exist measures $\nu_n \in M(G)$ such that $\|\nu_n\| \leq K$ and $\int_G f\, d\nu_n = T_n(f)$ for all $f \in C(G)$. If $\psi \in E \subset C_E(G)$, then

$$\hat{\nu}_n(\psi^{-1}) = \int_G \psi\, d\nu_n = T_n(\psi) = \begin{cases} 1 & \text{if } \psi \in X_n \\ 0 & \text{if } \psi \in X \setminus X_n. \end{cases}$$

This shows that (1) - (3) hold with $\mu_n(B) = \nu_n(-B)$ for Borel sets B.

Now suppose that $\{\mu_n\}_{n=1}^{\infty}$ satisfies (1) - (3). Then for each $f \in C_E(G)$ we have

$$\|s_n f\|_u = \|f * \mu_n\|_u \leq K \|f\|_u. \tag{4}$$

Consider f in $C_E(G)$ and $\varepsilon > 0$. There exists $p \in \mathrm{Trig}^+(G)$ such that $\|p\|_1 = 1$ and $\|f*p - f\|_u < \varepsilon/(K+1)$. For all n sufficiently large we have $X_n \supset \mathrm{Supp}(\hat{p})$ and so by (4) above,

$$\begin{aligned} \|s_n f - f\|_u &\leq \|s_n f - f*p\|_u + \|f*p - f\|_u \\ &= \|s_n(f - f*p)\|_u + \|f*p - f\|_u \\ &\leq (K+1)\|f - f*p\|_u < \varepsilon. \quad \square \end{aligned}$$

*6.14 THEOREM. <u>If</u> G <u>is infinite and</u> X <u>is countable, there exists a UC-set</u> E <u>in</u> X <u>that is not a Sidon set</u>.

<u>Proof</u>. A simple modification of the proof of Theorem 2.8 shows that there is an increasing sequence $\{n_m\}_{m=1}^{\infty}$ in $\mathbb{Z}^+$ and a dissociate set $D = \{\chi_k\}_{k=1}^{\infty}$ in X such that

$$\{\chi_1,\ldots,\chi_m\} \cup \{\chi_j\chi_k : 1 \leqq j \leqq m, 1 \leqq k \leqq m\} \subset X_{n_m} \tag{1}$$

and

$$\chi_{m+1} \notin X_{n_m}^{-1} X_{n_m}. \tag{2}$$

Let $A = \{\chi_j : j$ is odd$\}$, $B = \{\chi_k : k$ is even$\}$ and $E = AB$. Corollary 6.6, alias Theorem 1.4, shows that E is not a Sidon set.

We prove that E is a UC-set by establishing (1) - (3) of Lemma 6.13. Consider any n in $\mathbb{Z}^+$ with $n \geqq n_1$ and choose m so that $n_m \leqq n < n_{m+1}$. Observe that $\chi_j\chi_k$ belongs to X_{n_m} whenever $\max\{j,k\} \leqq m$ by (1). Also, if $j \neq k$ and $\chi_j\chi_k$ is in $X_{n_{m+1}} \setminus X_{n_m}$, then $\max\{j,k\} = m+1$, since otherwise we would have $j < k$, say, where $k > m+1$ so that

$$\chi_k \in \chi_j^{-1} X_{n_{m+1}} \subset X_{n_j}^{-1} X_{n_{m+1}} \subset X_{n_{k-1}}^{-1} X_{n_{k-1}},$$

which contradicts (2). These observations show that

$$E \cap X_n = \{\chi_j\chi_k : j+k \text{ odd}, 1 \leqq j \leqq m, 1 \leqq k \leqq m\} \cup C\chi_{m+1} \tag{3}$$

where $C \subset \{\chi_1,\ldots,\chi_m\}$ and the union is disjoint. Applying 2.7 to $D_m = \{\chi_1,\ldots,\chi_m\}$ and $\phi = \frac{1}{2}$, we obtain $\nu_n \in M^+(G)$ such that $\|\nu_n\| = 4$,

$$\hat{\nu}_n(\chi_j\chi_k) = 1 \quad \text{for } j \neq k, 1 \leqq j \leqq m, 1 \leqq k \leqq m, \tag{4}$$

and

$$\psi \in \operatorname{Supp}(\hat{\nu}_n) \text{ implies } \psi = \prod_{\chi \in S} \chi \tag{5}$$

for some asymmetric set $S \subset D_m \cup D_m^{-1}$. Applying 2.7 to D_m and $\phi = \frac{1}{2}\xi_C$, we obtain $\sigma_n \in M^+(G)$ such that $\|\sigma_n\| = 2$, $\hat{\sigma}_n(\chi_j) = 1$ for $\chi_j \in C$, and $\hat{\sigma}_n(\chi_j) = 0$ for $\chi_j \in D_m \setminus C$; also

(5) holds for σ_n. If $\tau_n = \chi_{m+1}\sigma_n$, then $\|\tau_n\| = 2$,

$$\hat{\tau}_n(\chi_{m+1}\chi_j) = \begin{cases} 1 & \text{for } \chi_j \in C \\ 0 & \text{for } \chi_j \in D_m \setminus C, \end{cases} \tag{6}$$

and

$$\psi \in \mathrm{Supp}(\hat{\tau}_n) \text{ implies } \psi = \chi_{m+1} \prod_{\chi \in T} \chi \tag{7}$$

for some asymmetric set $T \subset D_m \cup D_m^{-1}$. Now let $\mu_n = \nu_n + \tau_n$; clearly 6.13(1) holds with $K = 6$.

We next show that

$$\mathrm{Supp}(\hat{\nu}_n) \cap E = \{\chi_j\chi_k : j+k \text{ odd}, \ 1 \leqq j \leqq m, \ 1 \leqq k \leqq m\} \tag{8}$$

and

$$\mathrm{Supp}(\hat{\tau}_n) \cap E = C\chi_{m+1}. \tag{9}$$

If (8) fails, then E contains an element $\psi = \chi_j\chi_k$ where $j \neq k$ and $\max\{j,k\} \geqq m+1$ which, by (5), also has the form $\prod_{\chi \in S} \chi$ where S is an asymmetric subset of $D_m \cup D_m^{-1}$. Then

$$\chi_j \chi_k \prod_{\chi \in S} \chi^{-1} = 1$$

and, since D is dissociate, $\chi_{\max\{j,k\}} = 1$ which is impossible. Since $\hat{\tau}_n(\chi_{m+1}\chi_j) = 0$ for $\chi_j \in D_m \setminus C$ by (6), the failure of (9) would imply that $E \cap \mathrm{Supp}(\hat{\tau}_n)$ contains a character $\psi = \chi_j\chi_k$ where $j \neq k$ and $\max\{j,k\} \neq m+1$. By (7) we have $\psi = \chi_{m+1}\prod_{\chi \in T} \chi$ where T is an asymmetric subset of $D_m \cup D_m^{-1}$. Then we have $\chi_{m+1}\chi_j^{-1}\chi_k^{-1}\prod_{\chi \in T} \chi = 1$. Since D is dissociate, we conclude that

$$\text{if } \max\{j,k\} \leqq m, \text{ then } \chi_{m+1} = 1$$

and

$$\text{if } \max\{j,k\} \geqq m+2, \text{ then } \chi_{\max\{j,k\}} = 1.$$

Either way we have a contradiction. Since $\mu_n = \nu_n + \tau_n$, relations (4), (6), (8) and (9) together yield (2) and (3) of Lemma 6.13 for $n \geq n_1$. Therefore E is a UC-set for the sequence $\{X_n : n \geq n_1\}$ and hence for $\{X_n\}_{n=1}^{\infty}$. □

More arithmetic properties of Sidon sets are obtained in Lemmas 8.8 and 8.9.

*6.15 REMARKS. Pedemonte [1974] proves that if $\{n_j\}_{j=1}^{\infty}$ is an increasing sequence of positive integers satisfying

$$n_j > n_1 + n_2 + \cdots + n_{j-1},$$

then each of the sets

$$E^s = \{n_{j_1} + n_{j_2} + \cdots + n_{j_s} : 1 \leq j_1 < j_2 < \cdots < j_s\}$$

is a UC-set in $\mathbb{Z}$, $s = 1,2,\ldots$. For $s > 1$, these sets are not Sidon sets. She also proves that UC-sets in $\mathbb{Z}$ cannot contain arbitrarily long arithmetic progressions. Similar results hold for the Cantor group $\mathbb{Z}(2)^{\aleph_0}$. It is not known whether the union of two UC-sets is another UC-set.

Dressler and Pigno [1974c] prove that if E is a UC-set in $\mathbb{Z}$ and if $\mu \in M(\mathbb{T})$ satisfies $\hat{\mu}(n) = 0$ for all $n > 0$ such that $n \notin E$, then $\mu \in M_0(\mathbb{T})$, i.e. $\hat{\mu} \in c_0(\mathbb{Z})$. For each $s = 2,3,\ldots,$ the set

$$T_s = \{a_1! + a_2! + \cdots + a_s! : 1 \leq a_1 < a_2 < \cdots < a_s\}$$

is a UC-set which is not a Sidon set. Moreover, if $\mu \in M(\mathbb{T})$ satisfies $\hat{\mu}(n)$ for all $n > 0$ such that $n \notin T_s$, then $\mu = f\,d\lambda$ for some $f \in L^1(\mathbb{T})$.

Chapter 7

FATOU-ZYGMUND PROPERTIES

Sidon sets were shown to have the Fatou-Zygmund property in Chapter 3; see Corollary 3.6 and Definition 2.2. We now study this property more closely and also investigate a localized version of this property. For a real-valued function f, f^+ will denote the function $\max(f,0)$; ξ_K denotes the characteristic function of the set K.

7.1 LEMMA. *Let* E *be a nonvoid symmetric subset of* X, K *a nonvoid compact subset of* G, *and* $\varkappa > 0$. *The following properties are equivalent*:

(i) *For all* $f \in \mathrm{Trig}_E^r(G)$,

$$\|f\|_A \leq \varkappa \|f^+ \xi_K\|_u. \tag{1}$$

(ii) *Every* f *in* $C_E^r(G)$ *belongs to* $A(G)$ *and satisfies* (1).

(iii) *Given* ϕ *in* $\ell_h^\infty(E)$, *there is* $\mu \in M^+(-K)$ *such that* $\hat{\mu}|_E = \phi$ *and* $\|\mu\| \leq \varkappa \|\phi\|_\infty$.

Proof. First we prove that (i) implies (ii). Let $\{h_\alpha\}$ be an approximate unit for $L^1(G)$ where $h_\alpha \in \mathrm{Trig}^+(G)$, $\hat{h}_\alpha \geq 0$ and $\|h_\alpha\|_1 = 1$ for all α ([HR; 28.53]). Consider f in $C_E^r(G)$ and note that

$$\lim_\alpha \|f * h_\alpha - f\|_u = 0. \tag{2}$$

Each $f * h_\alpha$ belongs to $\mathrm{Trig}_E^r(G)$ and so (i) implies

$$\|f * h_\alpha\|_A \leq \varkappa \|(f * h_\alpha)^+ \xi_K\|_u \quad \text{for all } \alpha. \tag{3}$$

Using the inequality $|a^+ - b^+| \leq |a - b|$, valid for $a, b \in \mathbb{R}$, we obtain

$$\left| \|(f*h_\alpha)^+ \xi_K\|_u - \|f^+ \xi_K\|_u \right| \leq \|(f*h_\alpha)^+ \xi_K - f^+ \xi_K\|_u$$
$$\leq \|(f*h_\alpha - f)\xi_K\|_u \leq \|f*h_\alpha - f\|_u$$

and hence

$$\lim_\alpha \|(f*h_\alpha)^+ \xi_K\|_u = \|f^+ \xi_K\|_u \tag{4}$$

by (2). Let $F \subset E$ be finite. Since $\lim_\alpha \hat{h}_\alpha(\chi) = 1$ for χ in X, we have

$$\sum_{\chi \in F} |\hat{f}(\chi)| = \lim_\alpha \sum_{\chi \in F} |\hat{f}(\chi)\hat{h}_\alpha(\chi)| \leq \lim\sup_\alpha \|f*h_\alpha\|_A$$

and so by (3) and (4)

$$\sum_{\chi \in F} |\hat{f}(\chi)| \leq \lim\sup_\alpha \varkappa \|(f*h_\alpha)^+ \xi_K\|_u = \varkappa \|f^+ \xi_K\|_u .$$

Since F is arbitrary, we see that f is in $A(G)$ and (1) holds for f. Thus (ii) holds.

We now prove that (ii) implies (iii). Let

$$C_E^r(K) = \{f|_K : f \in C_E^r(G)\}$$

and note that $C_E^r(K)$ is a real linear subspace of $C^r(K)$. For each g in $C_E^r(K)$, there is exactly one g' in $C_E^r(G)$ so that $g'|_K = g$; moreover g' is in $A(G)$. To see this, consider $f_1, f_2 \in C_E^r(G)$ such that $f_1|_K = f_2|_K$. By (1), we have

$$\|f_1 - f_2\|_A \leq \varkappa \|(f_1 - f_2)^+ \xi_K\|_u = 0$$

and so $f_1 = f_2$; also $f_1 \in A(G)$ by (ii).

Now consider ϕ in $\ell_h^\infty(E)$ and define $L : C_E^r(K) \to \mathbb{R}$ via $L(g) = \sum_{\chi \in E} \hat{g}'(\chi)\, \phi(\chi)$. The last paragraph shows that L is well defined, and L is clearly linear. For $g \in C_E^r(K)$, we use (1) to write

$$L(g) \leq |L(g)| \leq \|g'\|_A \|\phi\|_\infty \leq \varkappa \|g'^+ \xi_K\|_u \|\phi\|_\infty = \varkappa \|g^+\|_u \|\phi\|_\infty .$$

We now apply the Hahn-Banach theorem to the space $C^r(K)$ and the sublinear functional (or gauge) $p(g) = \varkappa \|\phi\|_\infty \|g^+\|_u$ on $C^r(K)$. Hence there is a linear map $L' : C^r(K) \to \mathbb{R}$ extending L and satisfying

$$L'(g) \leq \varkappa \|\phi\|_\infty \|g^+\|_u \quad \text{for all} \quad g \in C^r(K). \tag{5}$$

Applying (5) to both g and $-g$, we find that L' is bounded on $C^r(K)$ and $\|L'\| \leq \varkappa \|\phi\|_\infty$. By the Riesz representation theorem, there is $\nu \in M^r(K)$ satisfying $L'(g) = \int_K g\, d\nu$ for $g \in C^r(K)$; also $\|\nu\| \leq \varkappa \|\phi\|_\infty$. In fact, we have $\nu \in M^+(K)$ because $g \geq 0$ and (5) imply

$$-L'(g) = L'(-g) \leq \varkappa \|\phi\|_\infty \|(-g)^+\|_u = 0.$$

The desired measure $\mu \in M^+(-K)$ is defined by $\mu(B) = \nu(-B)$ for Borel sets $B \subset G$. To show $\hat{\mu}|_E = \phi$, consider $\chi_o \in E$. Then $(\mathrm{Re}\, \chi_o|_E)' = \mathrm{Re}\, \chi_o$ and $(\mathrm{Im}\, \chi_o|_E)' = \mathrm{Im}\, \chi_o$. Letting g be $\mathrm{Re}\, \chi_o|_E$ and $\mathrm{Im}\, \chi_o|_E$ in the relation

$$\sum_{\chi \in E} \hat{g}'(\chi)\phi(\chi) = \int_K g\, d\nu,$$

we conclude that

$$\hat{\mu}(\chi_o) = \hat{\nu}(\overline{\chi_o}) = \int_K \chi_o\, d\nu = \sum_{\chi \in E} \hat{\chi}_o(\chi)\phi(\chi) = \phi(\chi_o).$$

Finally, we prove that (iii) implies (i). Consider f in $\mathrm{Trig}_E^r(G)$ and choose $\phi \in \ell_h^\infty(E)$ so that $\phi \hat{f} = |\hat{f}|$ on E and $\|\phi\|_\infty = 1$. From (iii) we obtain $\mu \in M^+(-K)$ satisfying $\hat{\mu}|_E = \phi$ and $\|\mu\| \leq \varkappa$. Then we have

$$\|f\|_A = \sum_{\chi\in E} |\hat{f}(\chi)| = \sum_{\chi\in E} \hat{f}(\chi)\phi(\chi) = \sum_{\chi\in E} \hat{f}(\chi)\hat{\mu}(\chi) = f*\mu(0)$$

$$= \int_{-K} f(-x)d\mu(x) = \int_K f(x)d\mu(-x) \leq \int_K f^+(x)d\mu(-x)$$

$$\leq \|f^+ \xi_K\|_u \|\mu\| \leq \varkappa \|f^+ \xi_K\|_u$$

and so (1) holds. □

The next theorem is essentially due to Edwards, Hewitt and Ross [1972c; 8.7]; the treatment here is taken from López [1974]. The theorem is a descendent of earlier theorems due to Kahane [1957; page 310], Helson and Kahane [1965; Theorem 2] and Méla [1969; page 44]. By $\lim\sup_E \phi(\chi)$, we mean $\lim_F \sup_{E\setminus F}\phi(\chi)$ where the limit is taken over the net of finite subsets F of E. If E is finite and $\phi \geq 0$, we decree $\lim\sup_E \phi(\chi) = 0$.

7.2 THEOREM. Let E be a nonvoid symmetric subset of X and K a nonvoid compact subset of G. The following are equivalent:

(i) There are $\varkappa > 0$ and finite symmetric $F \subset E$ so that

$$f \in \mathrm{Trig}^r_{E\setminus F}(G) \text{ implies } \|f\|_A \leq \varkappa \|f^+ \xi_K\|_u. \qquad (1)$$

(ii) There are $\varkappa > 0$ and finite symmetric $F \subset E$ so that $f \in C^r_{E\setminus F}(G)$ implies $f \in A(G)$ and $\|f\|_A \leq \varkappa\|f^+ \xi_K\|_u$.

(iii) There are $\varkappa > 0$ and finite symmetric $F \subset E$ so that given $\phi \in \ell_h^\infty(E \setminus F)$, we have $\phi = \hat{\mu}|_{E\setminus F}$ where μ is in $M^+(-K)$ and $\|\mu\| \leq \varkappa\|\phi\|_\infty$.

(iv) There is finite symmetric $F \subset E$ so that given $\phi \in \ell_h^\infty(E \setminus F)$, we have $\phi = \hat{\mu}|_{E\setminus F}$ for some $\mu \in M^+(-K)$.

(v) <u>Given</u> $\phi \in \ell_h^\infty(E)$ <u>we have</u> $\lim \sup_E |\hat{\mu}(\chi) - \phi(\chi)| = 0$ <u>for some</u> $\mu \in M^+(-K)$.

(vi) <u>Given</u> $\phi \in \ell_h^\infty(E)$ <u>we have</u> $\lim \sup_E |\hat{\mu}(\chi) - \phi(\chi)| < 1$ <u>for some</u> $\mu \in M^+(-K)$.

(vii) <u>Given hermitian</u> $\phi : E \to \mathbb{T}$, <u>there is</u> $\mu \in M^+(-K)$ <u>satisfying</u> $\lim \sup_E |\hat{\mu}(\chi) - \phi(\chi)| < 1$.

<u>Proof</u>. The implications (i) $\Rightarrow$ (ii) $\Rightarrow$ (iii) follow from Lemma 7.1 and the implications (iii) $\Rightarrow$ (iv) $\Rightarrow$ (v) $\Rightarrow$ (vi) $\Rightarrow$ (vii) are all obvious.

To prove (vii) $\Rightarrow$ (i), we assume that (vii) holds while (i) fails. First we show that there is a sequence $\{F_n\}_{n=1}^\infty$ of pairwise disjoint finite symmetric subsets of E and a sequence $\{f_n\}_{n=1}^\infty$ in $\mathrm{Trig}_E^r(G)$ so that

$$\mathrm{Supp}(\hat{f}_n) = F_n, \quad \|f_n\|_A = 1 \quad \text{and} \quad \|f_n^+ \xi_K\|_u < 2^{-n} \tag{2}$$

for all n. Since (1) fails, there is $f_1 \in \mathrm{Trig}_E^r(G)$ where $\|f_1\|_A > 2\|f_1^+ \xi_K\|_u$. We may suppose that $\|f_1\|_A = 1$ so that $\|f_1^+ \xi_K\|_u < \frac{1}{2}$. Let $F_1 = \mathrm{Supp}(\hat{f}_1)$. Assume $F_1, \ldots, F_n$ and $f_1, \ldots, f_n$ have been chosen satisfying (2). Since (1) fails for $\varkappa = 2^{n+1}$ and $F = \bigcup_{k=1}^n F_k$, there is $f_{n+1} \in \mathrm{Trig}_E^r(G)$ so that $\mathrm{Supp}(\hat{f}_{n+1}) \cap (\bigcup_{k=1}^n F_k) = \emptyset$ and

$$1 = \|f_{n+1}\|_A > 2^{n+1} \|f_{n+1}^+ \xi_K\|_u.$$

We put $F_{n+1} = \mathrm{Supp}(\hat{f}_{n+1})$ to complete the inductive construction.

Now choose any hermitian $\phi : E \to \mathbb{T}$ satisfying

$$\hat{f}_n(\chi)\phi(\chi) = |\hat{f}_n(\chi)| \quad \text{for } \chi \in F_n. \tag{3}$$

Such a ϕ exists since the functions $\hat{f}_n$ and $|\hat{f}_n|$ are hermitian functions. By (vii), we can find $\mu \in M^+(-K)$, δ in $(0,1)$ and a finite symmetric set $F_0 \subset E$ such that

$$|\hat{\mu}(\chi) - \phi(\chi)| \leqq 1 - \delta \quad \text{for} \quad \chi \in E \setminus F_0. \tag{4}$$

Let $\nu \in M^+(K)$ be defined by $\nu(B) = \mu(-B)$ for Borel sets $B \subset G$. Choose N so that $n \geqq N$ implies $F_n \cap F_0 = \emptyset$. For $\chi \in F_n$, we have

$$\begin{aligned} \hat{f}_n(\chi)\hat{\mu}(\chi) &= \hat{f}_n(\chi)\phi(\chi) + \hat{f}_n(\chi)[\hat{\mu}(\chi) - \phi(\chi)] \\ &= |\hat{f}_n(\chi)| + \hat{f}_n(\chi)[\hat{\mu}(\chi) - \phi(\chi)] \end{aligned}$$

by (3), and so by (4) we have

$$\operatorname{Re} \hat{f}_n(\chi)\hat{\mu}(\chi) \geqq |\hat{f}_n(\chi)| - |\hat{f}_n(\chi)|(1 - \delta) = \delta|\hat{f}_n(\chi)|$$

provided $n \geqq N$. Hence we have

$$\begin{aligned} \int_K f_n^+ \, d\nu \geqq \int_K f_n \, d\nu &= \operatorname{Re} \int_K \Big[\sum_{\chi \in F_n} \hat{f}_n(\chi)\chi \Big] d\nu \\ &= \operatorname{Re} \sum_{\chi \in F_n} \hat{f}_n(\chi)\hat{\nu}(\chi^{-1}) = \sum_{\chi \in F_n} \operatorname{Re} \hat{f}_n(\chi)\hat{\mu}(\chi) \\ &\geqq \delta \sum_{\chi \in F_n} |\hat{f}_n(\chi)| = \delta \|f_n\|_A = \delta, \end{aligned}$$

by (2). Consequently, we have $\sum_{n \geqq N} \int_K f_n^+ \, d\nu = \infty$. On the other hand, (2) shows that

$$\sum_{n=1}^{\infty} \int_K f_n^+ \, d\nu \leqq \sum_{n=1}^{\infty} \int_K 2^{-n} \, d\nu = \nu(K) < \infty,$$

and this is clearly impossible. □

7.3 DEFINITION. Let K be a nonvoid compact subset of G. A symmetric set E in X satisfying any of the equivalent conditions in Theorem 7.2 will be called an <u>FZ(K)-set</u>. The set E will be called a <u>full FZ-set</u> provided it is an FZ(K)-set for all compact subsets K of G that have nonvoid interior.

This terminology is due to Edwards, Hewitt and Ross [1972c], but their definitions differ from the ones given above. However, we will show in 7.13 that the definitions are equivalent for countable sets E. Full FZ-sets are characterized in Theorem 8.16.

The reader will be relieved to learn that there is a close connection between the Fatou-Zygmund property defined in 2.2 and FZ(G)-sets.

7.4 PROPOSITION. <u>Let</u> E <u>be a symmetric set in</u> X <u>such that</u> $1 \notin E$. <u>Then</u> E <u>is an FZ(G)-set if and only if</u> E <u>satisfies the Fatou-Zygmund property</u>.

<u>Proof</u>. The Fatou-Zygmund property is stronger, so we need only suppose that E is an FZ(G)-set and obtain $\varkappa > 0$ such that if ϕ is in $\ell_h^{\infty}(E)$, then $\phi = \hat{\mu}|_E$ where $\mu \in M^+(G)$ and $\|\mu\| \leqq \varkappa \|\phi\|_{\infty}$. First we find $\varkappa$. By 7.2(iii), there are $\varkappa_1 > 0$ and finite symmetric $F \subset E$ so that if ϕ is in $\ell_h^{\infty}(E \setminus F)$, then

$$\phi = \hat{\mu}|_{E \setminus F} \quad \text{for some} \quad \mu \in M^+(G) \quad \text{with} \quad \|\mu\| \leqq \varkappa_1 \|\phi\|_{\infty}. \tag{1}$$

All norms on the finite-dimensional space $\ell_h^{\infty}(F)$ are equivalent and so $\varkappa_2 > 0$ exists satisfying

$$\|\phi\|_1 \leqq \varkappa_2 \|\phi\|_{\infty} \quad \text{for all} \quad \phi \in \ell_h^{\infty}(F). \tag{2}$$

Now let $\varkappa = \varkappa_1 + 2\varkappa_2 + 2\varkappa_1\varkappa_2$, and consider ϕ in $\ell_h^{\infty}(E)$. By (1) there is $\mu_1 \in M^+(G)$ satisfying

$$\|\mu_1\| \leqq \varkappa_1 \|\phi\|_{\infty} \quad \text{and} \quad \hat{\mu}_1(\chi) = \phi(\chi) \quad \text{for} \quad \chi \in E \setminus F. \tag{3}$$

Now let g be the member of $\mathrm{Trig}_F(G)$ satisfying

$$\hat{g}(\chi) = \phi(\chi) - \hat{\mu}_1(\chi) \quad \text{for} \quad \chi \in F. \tag{4}$$

Since $\hat{g}$ is hermitian, g is real valued. Let $\mu_2 = (g + \|g\|_u)\lambda$ and $\mu = \mu_1 + \mu_2 \in M^+(G)$. It is easy to see that $\hat{\mu}|_E = \phi$ since $\hat{\lambda}(\chi) = 0$ for $\chi \neq 1$. Using (2) - (4), we find

$$\|g\|_u \leqq \|g\|_A = \|\hat{g}\|_1 \leqq \varkappa_2 \|\hat{g}\|_\infty \leqq \varkappa_2(\|\phi\|_\infty + \|\mu_1\|)$$
$$\leqq \varkappa_2(\|\phi\|_\infty + \varkappa_1\|\phi\|_\infty) = (\varkappa_2 + \varkappa_1\varkappa_2)\|\phi\|_\infty$$

and hence

$$\|\mu\| \leqq \|\mu_1\| + \|g + \|g\|_u\|_1 \leqq \|\mu_1\| + 2\|g\|_u$$
$$\leqq \varkappa_1\|\phi\|_\infty + 2(\varkappa_2 + \varkappa_1\varkappa_2)\|\phi\|_\infty = \varkappa\|\phi\|_\infty. \quad \square$$

We now study several interesting properties of FZ(K)-sets. An idea of the sorts of properties that we will be investigating can be obtained by glancing at the statement of Theorem 7.11. The results in 7.6 - 7.11 are based on Edwards, Hewitt and Ross [1972c; §8]. The simplified treatment given below is taken from López [1974]. Moreover, Theorems 7.7 and 7.8 contain several improvements due to López.

We first prove a generalized open mapping theorem.

*7.5 LEMMA. <u>Let</u> $T : B_1 \to B_2$ <u>be a bounded linear transformation where</u> B_1 <u>and</u> B_2 <u>are Banach spaces. Suppose that</u> $\{A_n\}_{n=1}^\infty$ <u>is a sequence of subsets of</u> B_1 <u>satisfying</u>:

<u>each</u> A_n <u>is closed, convex, and bounded and</u> $0 \in A_n$, (1)

$$A_n + A_m \subset A_{n+m} \quad \text{for } m, n \in \mathbb{Z}^+, \tag{2}$$

$$r A_n \subset A_{rn} \quad \text{for } r, n \in \mathbb{Z}^+, \tag{3}$$

<u>and</u>

$$T\Big(\bigcup_{n=1}^\infty A_n\Big) = B_2. \tag{4}$$

<u>Then for some</u> $k \in \mathbb{Z}^+$ <u>we have</u>

$$\{y \in B_2 : \|y\| \leqq 1\} \subset T(A_k). \tag{5}$$

Proof. By the Baire category theorem, $T(A_{n_0})^-$ has non-void interior for some $n_0 \in \mathbb{Z}^+$. Hence some integer $r > 1$ and element $y_0 \in B_2$ satisfy

$$\{y \in B_2 : \|y - y_0\| \leqq \tfrac{1}{r}\} \subset T(A_{n_0})^-.$$

By (4) we have $-y_0 \in T(A_{n_1})$ for some $n_1 \in \mathbb{Z}^+$. Hence y in B_2 and $\|y\| \leqq 1/r$ imply

$$y = y + y_0 - y_0 \in T(A_{n_0})^- + T(A_{n_1}) \subset T(A_{n_0} + A_{n_1})^- \subset T(A_{n_0+n_1})^-.$$

It follows that for y in B_2 and $\|y\| \leqq 1$ we have

$$y \in rT(A_{n_0+n_1})^- \subset T(rA_{n_0+n_1})^- \subset T(A_{rn_0+rn_1})^-. \tag{6}$$

Let $C = A_{rn_0+rn_1}$ and $k = 2rn_0 + 2rn_1$. Then $2C \subset A_k$ by (3); and (6) can be rewritten as

$$y \in B_2 \text{ and } \|y\| \leqq 1 \text{ imply } y \in T(C)^-. \tag{7}$$

To prove (5) we consider $y \in B_2$ where $\|y\| \leqq 1$. By (7) there exists $x_0 \in C$ so that $\|y - T(x_0)\| \leqq \frac{1}{2}$. Suppose $x_0, x_1, \ldots, x_n$ have been selected in C so that

$$\|y - T(\sum_{i=0}^{n} 2^{-i} x_i)\| \leqq 2^{-n-1}. \tag{8}$$

Then by (7) there exists x_{n+1} in C satisfying

$$\|2^{n+1}[y - T(\sum_{i=0}^{n} 2^{-i} x_i)] - T(x_{n+1})\| \leqq \tfrac{1}{2}$$

so that

$$\|y - T(\sum_{i=0}^{n+1} 2^{-i} x_i)\| \leqq 2^{-n-2}.$$

Thus (8) holds for $n + 1$ and the inductive construction can continue. For $n = 0,1,2,\ldots,$ let $w_n = \sum_{i=0}^{n} 2^{-i} x_i$. Each w_n belongs to $2C \subset A_k$ since C is convex, $0 \in C$ and

$\Sigma_{i=0}^{n} 2^{-i} < 2$. The sequence $\{w_n\}_{n=0}^{\infty}$ is Cauchy in B_1 because $n > m$ implies

$$\|w_n - w_m\| \leq \sum_{i=m+1}^{n} 2^{-i} \|x_i\| \leq \sum_{i=m+1}^{n} 2^{-i}M$$

where $M = \sup\{\|x\| : x \in O\}$. Since A_k is closed and B_1 is complete, there exists $x \in A_k$ such that $x = \lim_n w_n$. Since T is continuous, (8) assures us that $T(x) = y$. This establishes (5). □

*7.6 REMARK. We will need the following fact:

$$\text{if } f \in L^{\infty}(G) \text{ and } \hat{f} \geq 0, \text{ then } f \in A(G). \tag{1}$$

To see this, consider f in $L^{\infty}(G)$ where $\hat{f} \geq 0$, and let $\{h_\alpha\}$ be an approximate unit for $L^1(G)$ where $h_\alpha \in \mathrm{Trig}^+(G)$, $\hat{h}_\alpha \geq 0$ and $\|h_\alpha\|_1 = 1$ for all α. For a finite subset F of X, we have

$$\sum_{\chi \in F} \hat{f}(\chi)\hat{h}_\alpha(\chi) \leq \sum_{\chi \in X} \hat{f}(\chi)\hat{h}_\alpha(\chi) = f*h_\alpha(0) \leq \|f*h_\alpha\|_\infty$$

$$\leq \|f\|_\infty \|h_\alpha\|_1 = \|f\|_\infty .$$

Taking the limit on α we obtain

$$\sum_{\chi \in F} \hat{f}(\chi) \leq \|f\|_\infty$$

for all finite subsets F of X and so

$$\|f\|_A = \sum_{\chi \in X} \hat{f}(\chi) \leq \|f\|_\infty < \infty .$$

This proves (1). Note that (1) was essentially proved in the proof of Theorem 1.3, (iii) ⇒ (v).

The matching properties 7.7(ii) - (v) that follow are akin to 7.2(iii) and 7.2(iv).

*7.7 THEOREM. *Let* E *be a nonvoid symmetric subset of* X *and* K *a compact subset of* G. *The following are equivalent:*

(i) E *is an FZ(K)-set.*

(ii) *There is a finite symmetric set* $F \subset E$ *so that given* $\phi \in \ell_h^\infty(E)$, *we have* $\phi = (g+\mu)\hat{}|_E$ *where* $\mu \in M^+(-K)$ *and* $g \in \mathrm{Trig}_F^r(G)$.

(iii) *There are* $\varkappa > 0$ *and a finite symmetric set* $F \subset E$ *so that given* $\phi \in \ell_h^\infty(E)$, *we have* $\phi = (g+\mu)\hat{}|_E$ *where* $\mu \in M^+(-K)$, $g \in \mathrm{Trig}_F^r(G)$ *and*

$$\|g\|_\infty + \|\mu\| \leqq \varkappa \|\phi\|_\infty. \tag{1}$$

(iv) *There exists* $\varkappa > 0$ *so that given* $\phi \in \ell_h^\infty(E)$, *we have* $\phi = (g+\mu)\hat{}|_E$ *where* $\mu \in M^+(-K)$, $g \in L_r^\infty(G)$ *and (1) holds.*

(v) *Given* $\phi \in \ell_h^\infty(E)$, *we have* $\phi = (g+\mu)\hat{}|_E$ *where* $\mu \in M^+(-K)$ *and* $g \in L_r^\infty(G)$.

(vi) *There are* $\varkappa > 0$ *and a finite symmetric set* $F \subset E$ *so that* $f \in L_{E\setminus F}^{\infty,r}(G)$ *implies* $f \in A(G)$ *and* $\|f\|_A \leqq \varkappa \|f^+ \xi_K\|_u$.

Proof. *(i) ⇒ (ii).* Recall that an FZ(K)-set satisfies all the properties in 7.2. By 7.2(iv) there is a finite symmetric set $F \subset E$ so that given $\phi \in \ell_h^\infty(E \setminus F)$, we have $\phi = \hat{\mu}|_{E\setminus F}$ for some $\mu \in M^+(-K)$. So, given $\phi \in \ell_h^\infty(E)$, we can write $\phi|_{E\setminus F} = \hat{\mu}|_{E\setminus F}$ for some $\mu \in M^+(-K)$. Clearly

$$g = \sum_{\chi \in F} [\phi(\chi) - \hat{\mu}(\chi)]\chi$$

belongs to $\mathrm{Trig}_F^r(G)$ and $(\hat{g} + \hat{\mu})|_E = \phi$. Thus (ii) holds.

(ii) ⇒ (iii). Let F be as in (ii) and let $B = \mathrm{Trig}_F^r(G) \times M^r(-K)$; this is a real Banach space with the norm $\|(g,\mu)\| = \|g\|_u + \|\mu\|$. Let $T : B \to \ell_h^\infty(E)$ be defined by $T(g,\mu) = (g+\mu)\hat{}|_E$; clearly T is a bounded linear transformation. For each n, let

$$A_n = \{(g,\mu) \in B : \mu \geqq 0 \text{ and } \|g\|_u + \|\mu\| \leqq n\}.$$

Then 7.5(1) - (3) obviously hold and 7.5(4) holds because F satisfies (ii). Hence for some $\varkappa \in \mathbb{Z}^+$ we have

$$\{\phi \in \ell_h^\infty(E) : \|\phi\|_\infty \leqq 1\} \subset T(A_\varkappa).$$

This inclusion implies (iii) directly.

<u>(iii) ⇒ (iv) ⇒(v)</u>. Obvious.

<u>(v) ⇒ (i)</u>. Consider ϕ in $\ell_h^\infty(E)$ so that $\phi = (g+\mu)\hat{}\,|_E$ where $\mu \in M^+(-K)$ and $g \in L_r^\infty(G)$. By the Riemann-Lebesgue lemma, we have $\hat{g} \in c_0(X)$ and so $\lim\sup_E |\phi(\chi) - \hat{\mu}(\chi)| = 0$. Thus E is an FZ(K)-set by 7.2(v).

<u>(i) ⇒ (vi)</u>. Let $\varkappa > 0$ and F be as in 7.2(iii), and consider f in $L^{\infty,r}_{E\setminus F}(G)$. There exists ϕ in $\ell_h^\infty(E\setminus F)$ such that $\phi(\chi)\hat{f}(\chi) = |\hat{f}(\chi)|$ for all $\chi \in E\setminus F$ and such that $\|\phi\|_\infty = 1$. By 7.2(iii) we can write $\phi = \hat{\mu}|_{E\setminus F}$ where μ is in $M^+(-K)$ and $\|\mu\| \leqq \varkappa$. Then $f*\mu$ belongs to $L_r^\infty(G)$ and $(f*\mu)\hat{} = |\hat{f}| \geqq 0$, and so $f*\mu$ belongs to $A(G)$ by 7.6(1). Moreover, we have

$$\|f\|_A = \sum_{\chi\in E\setminus F} |\hat{f}(\chi)| = \|f*\mu\|_A = f*\mu(0) = \int_{-K} f(-y)d\mu(y)$$

$$\leqq \int_{-K} f^+(-y)d\mu(y) \leqq \|f^+ \xi_K\|_u \|\mu\| \leqq \varkappa\|f^+ \xi_K\|_u.$$

<u>(vi) ⇒ (i)</u>. Clearly (vi) implies 7.2(ii). □

*7.8 THEOREM. <u>Let</u> E <u>be an FZ(K)-set for some compact subset</u> K <u>of</u> G. <u>There exists</u> $\varkappa > 0$ <u>and a finite symmetric subset</u> F <u>of</u> E <u>with the following property. If</u> V <u>is a closed neighborhood of</u> 0 <u>and if</u> f <u>in</u> $L^{1,r}_{E\setminus F}(G)$ <u>satisfies</u> $\|f^+ \xi_{K+V}\|_\infty < \infty$, <u>then</u> f <u>belongs to</u> $A(G)$ <u>and</u>

$$\|f\|_A \leqq \varkappa\|f^+ \xi_{K+V}\|_\infty. \tag{1}$$

In particular, if $f \in L^{1,r}_E(G)$ *and* $\|f^+ \xi_{K+V}\|_\infty < \infty$, *then* f *belongs to* $A(G)$.

Proof. Let $\varkappa > 0$ and F be as in 7.2(iii). Let V be a closed neighborhood of 0, which we may suppose is symmetric. Consider f in $L^{1,r}_{E\setminus F}(G)$ where $\|f^+ \xi_{K+V}\|_\infty < \infty$, and select $\phi \in \ell_h^\infty(E \setminus F)$ so that $\phi(\chi)\hat{f}(\chi) = |\hat{f}(\chi)|$ for $\chi \in E \setminus F$ and $\|\phi\|_\infty = 1$. By 7.2(iii), $\phi = \hat{\mu}|_{E\setminus F}$ where $\mu \in M^+(-K)$ and $\|\mu\| \leqq \varkappa$. Familiar arguments show that $L^1(G)$ has an approximate unit $\{k_\alpha\} \subset C^+(G)$ such that $\mathrm{Supp}(k_\alpha) \subset V$, $\|k_\alpha\|_1 = 1$ and $\hat{k}_\alpha \geqq 0$ for all α; compare [HR; 33.11]. Since each function $f*k_\alpha*\mu$ is in $C(G)$ and $(f*k_\alpha*\mu)^\wedge = |\hat{f}|\hat{k}_\alpha \geqq 0$, each $f*k_\alpha*\mu$ is in $A(G)$ by 7.6(1). For λ-almost all $y \in V$ we have

$$f*\mu(y) = \int_{-K} f(y-x)\,d\mu(x) \leqq \int_{-K} f^+(y-x)\,d\mu(x)$$

$$\leqq \|f^+ \xi_{K+V}\|_\infty \|\mu\| \leqq \varkappa \|f^+ \xi_{K+V}\|_\infty$$

and so

$$\sum_{\chi\in X} |\hat{f}(\chi)|\hat{k}_\alpha(\chi) = \sum_{\chi \in X} (f*k_\alpha*\mu)^\wedge(\chi) = f*k_\alpha*\mu(0)$$

$$= \int_V k_\alpha(-y) f*\mu(y)\,d\lambda(y) \leqq \varkappa\|f^+ \xi_{K+V}\|_\infty .$$

Since $\lim_\alpha \hat{k}_\alpha(\chi) = 1$ for all $\chi \in X$, it follows that $\|f\|_A \leqq \varkappa \|f^+ \xi_{K+V}\|_\infty$.

For the last sentence of the theorem, consider f in $L^{1,r}_E(G)$ and apply the preceding to $f - \sum_{\chi\in F} \hat{f}(\chi)\chi$. □

For a real-valued function f, ess sup(f) will denote its essential supremum. Similar remarks apply to sup(f) and max(f).

*7.9 COROLLARY. If $E \subset X$ is an FZ(G)-set and $1 \notin E$, then a constant $\varkappa > 0$ exists so that $f \in L_E^{1,r}(G)$ and $\operatorname{ess\,sup}(f) < \infty$ imply $f \in A(G)$ and $\|f\|_A \leq \varkappa \operatorname{ess\,sup}(f)$.

Proof. Since $1 \notin E$, we have $0 = \hat{f}(1) = \int_G f\,d\lambda$ and so $\operatorname{ess\,sup}(f) = \|f^+\|_\infty$. By Proposition 7.4, E satisfies the Fatou-Zygmund property defined in 2.2. So the proof of 7.8 applies with $F = \emptyset$ and $K = V = G$. □

In Theorem 7.11 we will summarize some properties equivalent to the Fatou-Zygmund property. First we prove an easy lemma.

*7.10 LEMMA. Let F be a finite symmetric set in X where $1 \notin F$ and let $\varepsilon > 0$. Then there is $f \in \operatorname{Trig}^+(G)$ such that $\hat{f}|_F = 1$ and $\|f\|_1 < 1 + \varepsilon$.

Proof. Let $\{h_\alpha\} \subset \operatorname{Trig}^+(G)$ be an approximate unit for $L^1(G)$ where $\|h_\alpha\|_1 = 1$ for all α and $\lim_\alpha \hat{h}_\alpha(\chi) = 1$ for all $\chi \in X$. For each α, the function

$$f_\alpha = h_\alpha + \sum_{\chi \in F} [1 - \hat{h}_\alpha(\chi)]\chi + \sum_{\chi \in F} |1 - \hat{h}_\alpha(\chi)| \cdot 1$$

belongs to $\operatorname{Trig}^+(G)$ and satisfies $\hat{f}_\alpha|_F = 1$. Also

$$\|f_\alpha\|_1 = \hat{f}_\alpha(1) = \hat{h}_\alpha(1) + \sum_{\chi \in F} |1 - \hat{h}_\alpha(\chi)|$$

and so $\lim_\alpha \|f_\alpha\|_1 = 1$. In particular, we have $\|f_\alpha\|_1 < 1 + \varepsilon$ for suitable α. □

*7.11 THEOREM. The following statements are equivalent for a symmetric set $E \subset X$ such that $1 \notin E$.

(i) E is an FZ(G)-set; see 7.2.

(ii) E is a Sidon set.

(iii) There is $\varkappa > 0$ so that each ϕ in $\ell_h^\infty(E)$ has the form $\hat{\mu}|_E$ where $\mu \in M^+(G)$ and $\|\mu\| \leq \varkappa \|\phi\|_\infty$.

(iv) Each ϕ in $\ell_h^\infty(E)$ has the form $\hat{\mu}|_E$ for some μ in $M^+(G)$.

(v) There is $\kappa > 0$ so that each ϕ in $c_{oh}(E)$ has the form $\hat{f}|_E$ where $f \in L_+^1(G)$ and $\|f\|_1 \leqq \kappa\|\phi\|_\infty$.

(vi) Each ϕ in $c_{oh}(E)$ has the form $\hat{f}|_E$ for some f in $L_+^1(G)$.

(vii) There is $\kappa > 0$ so that $\|f\|_A \leqq \kappa \max(f)$ for all f in $\mathrm{Trig}_E^r(G)$.

(viii) There is $\kappa > 0$ so that $\|f\|_A \leqq \kappa \operatorname{ess\,sup}(f)$ for all f in $L_E^{1,r}(G)$.

(ix) If $f \in L_E^{1,r}(G)$ and $\operatorname{ess\,sup}(f) < \infty$, then $f \in A(G)$.

(x) If $\mu \in M_E^r(G)$ and $\mu \leqq c\lambda$ for some $c \in \mathbb{R}$, then $\mu = f\lambda$ where $f \in A(G)$.

(xi) There is $\kappa > 0$ so that $\mu \in M_E^r(G)$ and $\mu \leqq c\lambda$ for some $c \in \mathbb{R}$ imply $\mu = f\lambda$ where $f \in A(G)$ and $\|f\|_A \leqq \kappa c$.

Proof. The equivalence of (i) and (iii) is proved in 7.4 and the equivalence of (ii) and (iii) follows from Drury's Theorem 3.3. We will show

$$\text{(iii)} \Rightarrow \text{(v)} \Rightarrow \text{(vi)} \Rightarrow \text{(iv)} \Rightarrow \text{(iii)}$$

and

$$\text{(iii)} \Rightarrow \text{(xi)} \Rightarrow \text{(x)} \Rightarrow \text{(ix)} \Rightarrow \text{(viii)} \Rightarrow \text{(vii)} \Rightarrow \text{(i)}.$$

(iii) $\Rightarrow$ (v). Let κ be as in (iii). Consider ϕ in $c_{oh}(E)$; we may suppose that $\|\phi\|_\infty = 1$. For each $n \in \mathbb{Z}^+$, let

$$E_n = \{\chi \in E : 2^{-n} < |\phi(\chi)| \leqq 2^{-n+1}\}.$$

Note that each E_n is symmetric and finite. For each n, let $\phi_n = \phi\xi_{E_n} \in \ell_h^\infty(E)$ and use (iii) to obtain $\mu_n \in M^+(G)$ so that $\hat{\mu}_n|_E = \phi_n$ and $\|\mu_n\| \leqq \kappa\|\phi_n\|_\infty \leqq 2^{-n+1}\kappa$. By Lemma 7.10, there is $f_n \in \mathrm{Trig}^+(G)$ so that $\hat{f}_n|_{E_n} = 1$ and $\|f_n\|_1 < 2$. For each n we have

$$\|\mu_n * f_n\|_1 \leqq \|\mu_n\| \, \|f_n\|_1 \leqq 2^{-n+2}\varkappa.$$

Hence $f = \sum_{n=1}^{\infty} \mu_n * f_n$ defines a function in $L^1_+(G)$ for which $\|f\|_1 \leqq \sum_{n=1}^{\infty} 2^{-n+2}\varkappa = 4\varkappa = 4\varkappa\|\phi\|_\infty$. Also we have

$$\hat{f}|_E = \sum_{n=1}^{\infty} \hat{\mu}_n|_E \hat{f}_n|_E = \sum_{n=1}^{\infty} \phi_n = \phi$$

and so (v) holds.

<u>(v) ⇒ (vi)</u>. Obvious.

<u>(vi) ⇒ (iv)</u>. An application of Lemma 7.5 to the mapping $T : L^1_r(G) \to c_{oh}(E)$ defined by $T(f) = \hat{f}|_E$ and the sets $A_n = \{f \in L^1_+(G) : \|f\|_1 \leqq n\}$ shows that (v) holds for some $\varkappa$. Let ϕ be in $\ell^\infty_h(E)$. Then to each finite symmetric subset F of E there corresponds $f_F \in L^1_+(G)$ such that $\hat{f}_F|_E = \phi\xi_F$ and $\|f_F\|_1 \leqq \varkappa\|\phi\|_\infty$. Since $\{f_F : F \subset E$ is finite symmetric$\}$ is a net in $\{\nu \in M^+(G) : \|\nu\| \leqq \varkappa\|\phi\|_\infty\}$, Alaoglu's theorem shows that some $\mu \in M^+(G)$ is a weak-* cluster point of $\{f_F\}$. It is easy to check that $\hat{\mu}|_E = \phi$.

<u>(iv) ⇒ (iii)</u>. This is another straightforward application of Lemma 7.5.

<u>(iii) ⇒ (xi)</u>. Let $\varkappa$ be as in (iii) and suppose that μ is in $M^r_E(G)$ and $\mu \leqq c\lambda$ for some $c \in \mathbb{R}$. Since $0 = \hat{\mu}(1) = \mu(G) \leqq c\lambda(G) = c$, we see that $c \geqq 0$. In view of (iii), we can find $\nu \in M^+(G)$ so that $\hat{\nu}\hat{\mu} = |\hat{\mu}| \geqq 0$ and $\|\nu\| \leqq \varkappa$. Let $\{h_\alpha\} \subset \mathrm{Trig}^+(G)$ be an approximate unit for $L^1(G)$ where $\|h_\alpha\|_1 = 1$ for all α. For each finite set $F \subset X$ we have

$$\sum_{\chi\in F} |\hat{\mu}(\chi)|\hat{h}_\alpha(\chi) \leqq \sum_{\chi\in X} (\nu*\mu*h_\alpha)\hat{\ }(\chi) = \nu*\mu*h_\alpha(0)$$
$$= \int_G \nu*h_\alpha(-y)d\mu(y) \leqq c\int_G \nu*h_\alpha(-y)d\lambda(y)$$
$$= c\|\nu*h_\alpha\|_1 \leqq c\|\nu\|\cdot\|h_\alpha\|_1 \leqq \kappa c.$$

Taking the limit on α, we obtain $\sum_{\chi\in F} |\hat{\mu}(\chi)| \leqq \kappa c$ and so $\|\hat{\mu}\|_1 \leqq \kappa c$. Hence $\mu = f\lambda$ where $f \in A(G)$ and $\|f\|_A \leqq \kappa c$.

<u>(xi) ⇒ (x)</u>. Obvious.

<u>(x) ⇒ (ix)</u>. Simply observe that if f is in $L_E^{1,r}(G)$ and $\mu = f\lambda$, then $\mu \leqq (\text{ess sup } f)\lambda$.

<u>(ix) ⇒ (viii)</u>. First we show that

$$\|f\|_1 \leqq 2\|f^+\|_u \quad \text{for} \quad f \in \mathrm{Trig}_E^r(G). \tag{1}$$

We have $f = f^+ - f^-$ where $f^+ = \max(f,0)$ and $f^- = -\min(f,0)$. Since $1 \notin E$, we have

$$0 = \hat{f}(1) = \int_G f\,d\lambda = \int_G f^+\,d\lambda - \int_G f^-\,d\lambda$$

and so

$$\|f\|_1 = \int_G (f^+ + f^-)d\lambda = 2\int_G f^+\,d\lambda \leqq 2\|f^+\|_u.$$

This establishes (1). Now assume that (viii) fails. Then Corollary 7.9 shows that E is not an FZ(G)-set. In particular, 7.2(i) fails for $K = G$ and so, for each finite symmetric set $F \subset E$ and $n \in \mathbb{Z}^+$, some f in $\mathrm{Trig}_{E\setminus F}^r(G)$ satisfies $\|f\|_A > 2^n\|f^+\|_u$. We can also arrange for $\|f\|_A = 1$. Hence an easy induction shows that there is a sequence $\{f_n\}_{n=1}^\infty$ in $\mathrm{Trig}_E^r(G)$ so that

$$\|f_n\|_A = 1 \quad \text{for all} \quad n \in \mathbb{Z}^+, \tag{2}$$
$$\|f_n^+\|_u < 2^{-n} \quad \text{for all} \quad n \in \mathbb{Z}^+, \tag{3}$$

and

$$\{\mathrm{Supp}(\hat{f}_n)\}_{n=1}^\infty \text{ is a pairwise disjoint family.} \tag{4}$$

Inequalities (1) and (3) show that $\|f_n\|_1 \leqq 2^{-n+1}$ for all n and so $f = \sum_{n=1}^{\infty} f_n$ defines an element in $L_E^{1,r}(G)$. From (3) we have

$$\operatorname{ess\,sup}(f) \leqq \| (\sum_{n=1}^{\infty} f_n)^+ \|_\infty \leqq \| \sum_{n=1}^{\infty} f_n^+ \|_\infty \leqq \sum_{n=1}^{\infty} \|f_n^+\|_\infty$$
$$< \sum_{n=1}^{\infty} 2^{-n} = 1 < \infty,$$

while from (4) and (2) we have $\|f\|_A = \sum_{n=1}^{\infty} \|f_n\|_A = \infty$. Hence (ix) fails if (viii) fails.

<u>(viii) ⇒ (vii)</u>. Obvious.

<u>(vii) ⇒ (i)</u>. Obviously (vii) implies 7.2(i) for $K = G$. □

Glicksberg [1974] shows that $L_+^1(G)$ in 7.11(vi) can be replaced by $L_+^1(G,\mu)$ for any μ in $M^+(G)$ such that $\hat{\mu}$ is in $c_0(X)$ and $\operatorname{Supp}(\mu) = G$. He also observes that a symmetric set $E \subset X$ without 1 is an FZ(G)-set if and only if

$$\mu \in M_{E\cup\{1\}}^+(G) \quad \text{implies} \quad \hat{\mu} \in \ell^1(X).$$

This condition is easily seen to be equivalent to 7.11(x).

*7.12 NOTATION. Suppose that $\{X_n\}_{n=1}^{\infty}$ is an increasing sequence of finite symmetric subsets of X, and let $X_\infty = \bigcup_{n=1}^{\infty} X_n$. As in 1.6, we write

$$s_n f = \sum_{\chi \in X_n} \hat{f}(\chi)\chi$$

for $f \in L^1(G)$. For any hermitian function ϕ defined on a symmetric subset E of X, we write $s_n\phi$ for

$$\sum_{\chi \in X_n \cap E} \phi(\chi)\chi.$$

Thus each $s_n\phi$ belongs to $\operatorname{Trig}_E^r(G)$.

Property 7.13(iii) below was the original definition of an FZ(K)-set given by Edwards, Hewitt and Ross [1972c].

*7.13 THEOREM. Let notation be as in 7.12. For a symmetric subset E of X_∞ and a compact subset K of G, the following are equivalent:

(i) E is an $FZ(K)$-set.

(ii) There are $\varkappa > 0$ and a finite symmetric set $F \subset E$ so that

$$f \in \mathrm{Trig}^r_{E\setminus F}(G) \text{ implies } \|f\|_A \leqq \varkappa \sup_n \|(s_n f)^+ \xi_K\|_u. \quad (1)$$

(iii) For each hermitian function ϕ on E,

$$\sup_n \|(s_n\phi)^+ \xi_K\|_u < \infty \text{ implies } \phi \in \ell^1(E). \quad (2)$$

Proof. (i) ⇒ (ii). If f is in $\mathrm{Trig}_E(G)$, then we have $f = s_n f$ for large n and so 7.2(i) clearly implies (ii).

(ii) ⇒ (iii). Let $\varkappa$ and F be as in (ii) and consider a hermitian function ϕ on E for which

$$\sup_n \|(s_n\phi)^+ \xi_K\|_u < \infty. \quad (3)$$

Let $\phi_1 = \phi\xi_{E\setminus F}$ and $\phi_2 = -\phi\xi_F$. Clearly (3) holds for ϕ_2 and so (3) also holds for $\phi_1 = \phi + \phi_2$. Since it suffices to show that $\phi_1 \in \ell^1(E)$, we may suppose that ϕ vanishes on F. Let $M = \sup_n \|(s_n\phi)^+ \xi_K\|_u$. Each $s_n\phi$ belongs to $\mathrm{Trig}^r_{E\setminus F}(G)$ and so (1) shows that

$$\|s_n\phi\|_A \leqq \varkappa \sup_m \|(s_m s_n\phi)^+ \xi_K\|_u \leqq \varkappa M$$

since $s_m s_n \phi = s_{\min(m,n)}\phi$. Thus $\Sigma_{\chi\in X_n\cap E} |\phi(\chi)| \leqq \varkappa M$ for all n. Since $E \subset \bigcup_{n=1}^{\infty} X_n$, we infer that $\|\phi\|_1 \leqq \varkappa M$ and $\phi \in \ell^1(E)$.

(iii) ⇒ (i). Suppose that (i) does not hold; we will show that (iii) does not hold. A familiar construction yields a sequence $\{f_j\}_{j=1}^{\infty}$ in $\mathrm{Trig}^r_E(G)$ and an increasing sequence

$\{n_j\}_{j=1}^{\infty}$ in $\mathbb{Z}^+$ satisfying

$$\|f_j\|_A = 1 \quad \text{for all } j, \tag{4}$$

$$\|f_j^+ \xi_K\|_u \leqq 2^{-j} \quad \text{for all } j, \tag{5}$$

and

$$\operatorname{Supp}(\hat{f}_j) \subset E \setminus X_{n_{j-1}} \quad \text{and} \quad \operatorname{Supp}(\hat{f}_j) \subset X_{n_j} \quad \text{for } j \geqq 1, \tag{6}$$

where $n_0 = 1$; compare 1.7(2), 1.7(3) and 1.7(5). The sets $\operatorname{Supp}(\hat{f}_j)$ are pairwise disjoint by (6) and so $\phi = \Sigma_{j=1}^{\infty} \hat{f}_j|_E$ defines a hermitian function on E for which

$$\|\phi\|_1 = \sum_{j=1}^{\infty} \|f_j\|_A = \infty.$$

We will show that

$$\sup_n \|(s_n\phi)^+ \xi_K\|_u \leqq 2, \tag{7}$$

so that (iii) does not hold. We have $s_{n_0}\phi = 0$ and $s_{n_k}\phi = \Sigma_{j=1}^{k} f_j$ for $k \geqq 1$ so that

$$\|(s_{n_k}\phi)^+ \xi_K\|_u = \|(\sum_{j=1}^{k} f_j)^+ \xi_K\|_u \leqq \sum_{j=1}^{k} \|f_j^+ \xi_K\|_u < 1 \tag{8}$$

by (5). For any $n \in \mathbb{Z}^+$ we have $n_k \leqq n < n_{k+1}$ for some k; and (6) implies that

$$\|(s_n\phi - s_{n_k}\phi)^+ \xi_K\|_u \leqq \|s_n\phi - s_{n_k}\phi\|_A \leqq \|f_{k+1}\|_A = 1. \tag{9}$$

Inequalities (8) and (9) show that $\|(s_n\phi)^+ \xi_K\|_u < 2$ for each n and so (7) holds. □

*7.14 COROLLARY. _The following are equivalent for a symmetric subset_ E _of_ X_∞ _such that_ $1 \notin E$.

(i) E _is an_ _FZ(G)-set_.

(ii) _There is a constant_ $\varkappa > 0$ _so that_

$$f \in \operatorname{Trig}_E^r(G) \ \ \underline{\text{implies}} \ \ \|f\|_A \leqq \varkappa \sup_n \|(s_n f)^+\|_u. \tag{1}$$

(iii) For each hermitian function ϕ on E,

$$\sup_n \|(s_n\phi)^+\|_u < \infty \text{ implies } \phi \in \ell^1(E). \qquad (2)$$

Proof. Clearly 7.11(vii) implies (1) and so (i) implies (ii). The remaining implications follow from 7.13. □

The next theorem is an analogue of Theorem 7.13. Its proof is the same: simply remove the + signs from the proof of 7.13.

*7.15 THEOREM. Again notation is as in 7.12. The following are equivalent for a subset E of X_∞ and a compact subset K of G.

(i) There are $\varkappa > 0$ and a finite set $F \subset E$ such that

$$f \in \mathrm{Trig}_{E\setminus F}(G) \text{ implies } \|f\|_A \leq \varkappa \|f\, \xi_K\|_u.$$

(ii) There are $\varkappa > 0$ and a finite set $F \subset E$ so that

$$f \in \mathrm{Trig}_{E\setminus F}(G) \text{ implies } \|f\|_A \leq \varkappa \sup_n \|(s_n f)\, \xi_K\|_u.$$

(iii) For each complex-valued function ϕ on E,

$$\sup_n \|(s_n\phi)\, \xi_K\|_u < \infty \text{ implies } \phi \in \ell^1(E).$$

*7.16 COROLLARY. The following are equivalent for any subset E of X_∞.

(i) E is a Sidon set.

(ii) There is a constant $\varkappa > 0$ so that $f \in \mathrm{Trig}_E(G)$ implies $\|f\|_A \leq \varkappa \sup_n \|s_n f\|_u$.

(iii) For each complex-valued function ϕ on E,

$$\sup_n \|s_n\phi\|_u < \infty \text{ implies } \phi \in \ell^1(E).$$

Improvements of Corollaries 7.14 and 7.16 will be obtained in Theorems 9.15 and 9.17.

Chapter 8

DÉCHAMPS-GONDIM'S THEOREMS

8.1 DEFINITION. Let K be a nonvoid compact subset of G and E a subset of X. We say that E and K are <u>strictly associated</u> if there is a constant $\varkappa > 0$ such that

$$\|f\|_A \leqq \varkappa \|f \xi_K\|_u \quad \text{for all} \quad f \in \mathrm{Trig}_E(G). \tag{1}$$

We say that E and K are <u>associated</u> if $E \setminus F$ and K are strictly associated for some finite subset F of E.

Note that a Sidon set E is automatically strictly associated with G by Theorem 1.3(vii). Much more is true. Déchamps-Gondim [1970a] announced the following theorem: (I) <u>If E is a Sidon set in the character group X of a compact connected abelian group G, then E is associated with every compact subset K of G having nonvoid interior;</u> see Corollary 8.19. Subsequently [1970b] she announced the following improvement: (II) <u>If E is a Sidon set in the character group X of a compact connected abelian group G, then E is strictly associated with every compact subset K of G having nonvoid interior;</u> see Corollary 8.23. The proof of (II) relies on that of (I). The details of these results were presented in Déchamps-Gondim [1972]. Analogous results hold without the connectedness hypothesis; they involve the following notions.

8.2 DEFINITION. Let X_0 denote a subgroup of an abelian group X. A subset E of X is called X_0-*subtransversal* provided each coset of X_0 intersects E in at most one point. We say that E is *almost* X_0-*subtransversal* if E is the union of a finite set and an X_0-subtransversal set.

Using Déchamps-Gondim's methods, Ross [1972] showed the following: (I') *A Sidon set* E *is associated with every compact subset of* G *with nonvoid interior if and only if* E *is almost* X_0-*subtransversal for all finite subgroups* X_0 *of* X. (II') *A Sidon set* E *is strictly associated with every compact subset of* G *with nonvoid interior if and only if* E *is* X_0-*subtransversal for all finite subgroups* X_0 *of* X. See Corollary 8.18 and Theorem 8.22. Ross [1973] used the methods that established (I') to prove: *A symmetric* $FZ(G)$-*set* E *is a full* FZ-*set if and only if* E *is almost* X_0-*subtransversal for all finite subgroups* X_0 *of* X. This result now holds for symmetric Sidon sets thanks to Drury's Theorem 3.3; see Theorem 8.16. No viable analogue to (II') seems possible for FZ-sets.

First we prove the trivial halves of these results.

8.3 LEMMA. *Let* E *be a subset of* X.

(i) *If* E *is associated with every compact set* K *in* G *with nonvoid interior or if* E *is a (symmetric) full* FZ-*set, then* E *is almost* X_0-*subtransversal for all finite subgroups* X_0 *of* X.

(ii) *If* E *is strictly associated with every compact set* K *in* G *with nonvoid interior, then* E *is* X_0-*subtransversal for all finite subgroups* X_0 *of* X.

[Note that if G is connected, then X is torsion-free and $\{1\}$ is its only finite subgroup. In this case, the conclusions of this lemma are banal. Note also that E is X_0-subtransversal for all finite subgroups X_0 of X if and only if $\chi\psi^{-1}$ has infinite order for every two distinct members χ, ψ of E.]

<u>Proof</u>. <u>(ii)</u> Let X_0 be a finite subgroup of X. The annihilator G_0 of X_0 in G is an open subgroup, and so E and G_0 are strictly associated. Hence there exists $\varkappa > 0$ satisfying

$$\|f\|_A \leqq \varkappa \|f\, \xi_{G_0}\|_u \quad \text{for all} \quad f \in \mathrm{Trig}_E(G). \tag{1}$$

If χ and ψ are in E and $\chi\psi^{-1} \in X_0$, then $(\chi - \psi)\xi_{G_0} = 0$ and so (1) shows that $\chi = \psi$. Thus E is X_0-subtransversal.

<u>(i)</u> Suppose that E is associated with every compact set K with nonvoid interior. Let X_0 be a finite subgroup of X and let G_0 be its annihilator. Then there is a finite subset F of E so that $E \setminus F$ and G_0 are strictly associated. Thus (1) holds for $E \setminus F$, and it follows that $E \setminus F$ is X_0-subtransversal. I.e. E is almost X_0-subtransversal.

It remains to observe that every full FZ-set E is associated with every compact set K with nonvoid interior. Since E is an FZ(K)-set, there are $\varkappa > 0$ and finite symmetric $F \subset E$ so that $\|f\|_A \leqq \varkappa \|f^+ \xi_K\|_u$ for $f \in \mathrm{Trig}^r_{E\setminus F}(G)$. If f is in $\mathrm{Trig}_{E\setminus F}(G)$, this inequality applies to $\mathrm{Re}\, f$ and $\mathrm{Im}\, f$ to yield $\|f\|_A \leqq 2\varkappa \|f\, \xi_K\|_u$. Hence $E \setminus F$ and K are strictly associated and E and K are associated. □

The proof of the main theorem 8.16 is quite complicated and it seems worthwhile to first give a special case whose proof illustrates the general strategy. For this purpose, we introduce one more term.

8.4 DEFINITION. A set $E \subset X$ <u>tends to infinity</u> if given a finite subset Δ of X there exists a finite subset F of E such that

$$\chi, \psi \in E \setminus F \text{ and } \chi \neq \psi \text{ imply } \chi\psi^{-1} \notin \Delta. \qquad (1)$$

8.5 REMARKS. (a) A set $E \subset \mathbb{Z}$ tends to infinity if given $m \in \mathbb{Z}^+$ there exists $n \in \mathbb{Z}^+$ such that

$$j,k \in E, \quad |j| \geqq n, \quad |k| \geqq n \quad \text{and} \quad j \neq k \quad \text{imply} \quad |j-k| > m.$$

(b) It is trivial to verify that if a set $E \subset X$ tends to infinity, then E is almost X_0-subtransversal for all finite subgroups X_0 of X. The converse can fail: If G is connected, for example, then X is torsion-free and every subset of X is vacuously almost X_0-subtransversal for all finite subgroups X_0 of X.

(c) If G is 0-dimensional, then X is a torsion group and it is easy to see that a set $E \subset X$ tends to infinity if and only if E is almost X_0-subtransversal for all finite subgroups X_0 of X. In this case, full FZ-sets tend to infinity by 8.3(i). Note also that if E is X_0-subtransversal for all finite subgroups X_0 of X, then E contains at most one element.

(d) If E_0 is a dissociate set, then $E = E_0 \cup E_0^{-1}$ tends to infinity. To see this, we first show that

$$\{\chi \in E : \chi^2 = \phi\} \text{ has at most two elements for } \phi \in X \setminus \{1\}. \qquad (1)$$

If (1) fails, there exist distinct $\chi_1, \chi_2 \in E$ so that $\chi_1 \neq \chi_2^{-1}$ and $\chi_1^2 = \chi_2^2 \neq 1$. There exist $\phi_1, \phi_2 \in E_0$ and $\delta_1, \delta_2 \in \{-1,1\}$ so that $\chi_j^{\delta_j} = \phi_j$ for $j = 1,2$. Then $\phi_1 \neq \phi_2$ and $\phi_1^{2\delta_1}\phi_2^{-2\delta_2} = 1$. Since $\phi_1^{2\delta_1} \neq 1$, this contradicts the fact that E_0 is dissociate.

Now assume that E does not tend to infinity. Then X contains a finite set Δ with the following property. If F is a finite subset of E, then $E \setminus F$ contains distinct elements χ, ψ such that $\chi\psi^{-1} \in \Delta$. It follows that E contains sequences $\{\chi_n\}_{n=1}^{\infty}$ and $\{\psi_n\}_{n=1}^{\infty}$ such that $\chi_n\psi_n^{-1} \in \Delta \setminus \{1\}$ for all n and such that none of $\chi_n, \psi_n, \chi_n^{-1}, \psi_n^{-1}$ belong to $\{\chi_1, \psi_1, \ldots, \chi_{n-1}, \psi_{n-1}\}$ for $n \geqq 2$. Since Δ is finite we can, by passing to a subsequence if necessary, assume that $\chi_n\psi_n^{-1} = \chi_1\psi_1^{-1}$ for all n. In view of (1), we may also assume that $\chi_n^2 \notin \Delta$ for all n. There exist $\phi_j \in E_0$ and $\delta_j \in \{-1,1\}$ for $j = 1,2,3,4$ so that $\phi_1^{\delta_1} = \chi_1$, $\phi_2^{\delta_2} = \psi_1$, $\phi_3^{\delta_3} = \chi_2$ and $\phi_4^{\delta_4} = \psi_2$. The characters $\phi_1, \phi_2, \phi_3, \phi_4$ are distinct, as can be easily shown. If, for example, $\phi_1 = \phi_2$, then we would have $\chi_1 = \psi_1^{-1}$ and so $\chi_1^2 = \chi_1\psi_1^{-1} \in \Delta$, a contradiction. Since

$$\phi_1^{\delta_1}\phi_2^{-\delta_2}\phi_3^{-\delta_3}\phi_4^{\delta_4} = 1,$$

we again contradict the fact that E_0 is dissociate.

*(e) Many Rider sets tend to infinity. In fact, a Rider set E tends to infinity if and only if

$$\{\chi \in E \cap E^{-1} : \chi^2 = \phi\} \text{ is finite for all } \phi \in X \setminus \{1\}. \qquad (2)$$

Note that (1) is a special case of (2). Observe that (2) automatically holds in groups X, like $\mathbb{Z}$, that have no elements of order 2. Thus, in these groups, all Rider sets tend to infinity. Also observe that (2) vacuously holds for asymmetric sets E, so that asymmetric Rider sets always tend to infinity. A Rider set that does not tend to infinity is described in 8.25(f).

To see that a set E that tends to infinity must satisfy (2), consider $\phi \in X \setminus \{1\}$. By Definition 8.4, there is a finite subset F of E such that

$$\chi, \psi \in E \setminus F \text{ and } \chi \neq \psi \text{ imply } \chi\psi^{-1} \neq \phi. \tag{3}$$

If $\chi \in E \cap E^{-1}$ and $\chi^2 = \phi$, then $\chi, \chi^{-1} \in E$ and $\chi \neq \chi^{-1}$ and so either $\chi \in F$ or $\chi^{-1} \in F$ by (3). Thus we have

$$\{\chi \in E \cap E^{-1} : \chi^2 = \phi\} \subset F \cup F^{-1}$$

and so the set $\{\chi \in E \cap E^{-1} : \chi^2 = \phi\}$ is finite.

Now consider a Rider set E satisfying (2), and assume that E does not tend to infinity. As in part (d), E contains sequences $\{\chi_n\}_{n=1}^{\infty}$ and $\{\psi_n\}_{n=1}^{\infty}$ such that $\chi_n\psi_n^{-1} = \chi_1\psi_1^{-1} \equiv \phi_o \neq 1$ for all n and so that none of $\chi_n, \psi_n, \chi_n^{-1}, \psi_n^{-1}$ belong to $\{\chi_1, \psi_1, \ldots, \chi_{n-1}, \psi_{n-1}\}$ for $n \geqq 2$. In view of (2), we can also assume that

$$\chi_n \notin \{\chi \in E \cap E^{-1} : \chi^2 = \phi_o\} \text{ for all } n.$$

It follows that each set $\{\chi_1, \psi_1^{-1}, \chi_n^{-1}, \psi_n\}$ is an asymmetric subset of $E \cup E^{-1}$ consisting of four distinct characters and satisfying $\chi_1\psi_1^{-1}\chi_n^{-1}\psi_n = 1$. Hence $R_4(E,1) = \infty$ and E cannot be a Rider set; see 2.13.

Here is the promised special case of Theorem 8.16. It is essentially due to Déchamps-Gondim [1972; Théorème 5.1].

8.6 THEOREM. Every symmetric Sidon set $E \subset X$ which tends to infinity is a full FZ-set.

Proof. Let K be a closed subset of G with nonvoid interior; we show that E is an FZ(K)-set. Let g be a non-negative function in $A(G)$ such that

$$\operatorname{Supp}(g) \subset -K \quad \text{and} \quad \hat{g}(1) = \int_G g\,d\lambda = 1; \tag{1}$$

the existence of g follows from the regularity of the algebra $A(G)$ [R; 2.6.2]. Let $\varepsilon > 0$ be chosen so that $\varepsilon < 1$ and

$$4\varepsilon\,\|g\|_A < 1. \tag{2}$$

By Drury's Theorem 3.3 there is a constant $M > 0$ such that if ϕ_1 is a hermitian function on E and $\|\phi_1\|_\infty \leqq 1$, then there is $\nu \in M^+(G)$ satisfying

$$\hat{\nu}|_E = \phi_1, \quad \|\nu\| \leqq M, \quad \text{and} \quad |\hat{\nu}(\chi)| \leqq \varepsilon \quad \text{for} \quad \chi \notin E \cup \{1\}. \tag{3}$$

Now let $\Delta \subset X$ be a finite symmetric set satisfying

$$4M \sum_{\chi \in X \setminus \Delta} |\hat{g}(\chi)| < 1. \tag{4}$$

Since E tends to infinity, there is a finite symmetric $F \subset E$ such that

$$(E \setminus F)\cdot(E \setminus F) \cap \Delta \subset \{1\}. \tag{5}$$

By making F larger if necessary we also assume that

$$(E \setminus F) \cap \Delta = \emptyset. \tag{6}$$

Let $\phi : E \to \mathbb{T}$ be a hermitian function. By 7.2(vii) it suffices to find $\mu \in M^+(-K)$ so that

$$|\hat{\mu}(\chi) - \phi(\chi)| < \tfrac{1}{3} \quad \text{for} \quad \chi \in E \setminus F. \tag{7}$$

Define ϕ_1 on E by $\phi_1(\chi) = \phi(\chi)$ for $\chi \in E \setminus F$ and

$\phi_1(\chi) = 0$ for $\chi \in F$. Choose ν as in (3) and define $\mu = g\nu$. By (1) it is clear that $\mu \in M^+(-K)$.

We now show that (7) holds. Consider χ in $E \setminus F$ and note that

$$\hat{\mu}(\chi) = \int_G \chi^{-1} g \, d\nu = \sum_{\psi \in X} \hat{g}(\psi) \int_G \chi^{-1}\psi \, d\nu$$
$$= \sum_{\psi \in X} \hat{g}(\psi)\hat{\nu}(\chi\psi^{-1}) = \sum_{\psi \in X} \hat{g}(\chi\psi^{-1})\hat{\nu}(\psi).$$

Since $\hat{g}(1) = 1$ by (1) and $\hat{\nu}(\chi) = \phi(\chi)$ by (3), we can write

$$|\hat{\mu}(\chi) - \phi(\chi)| = \Big|\sum_{\psi \in X\setminus\{\chi\}} \hat{g}(\chi\psi^{-1})\hat{\nu}(\psi)\Big|$$
$$\leqq \sum_{\psi \in \chi\Delta\setminus\{\chi\}} |\hat{g}(\chi\psi^{-1})||\hat{\nu}(\psi)| + \sum_{\psi \in X\setminus\chi\Delta} |\hat{g}(\chi\psi^{-1})||\hat{\nu}(\psi)|$$
$$\equiv S_1 + S_2. \tag{8}$$

Note that by (5) we have $\chi^{-1}(E \setminus F) \cap \Delta \subset \{1\}$ and so $(E \setminus F) \cap \chi\Delta \subset \{\chi\}$. Thus if $\psi \in \chi\Delta \setminus \{\chi\}$, then $\psi \notin E \setminus F$. Also, (6) shows that if $\psi \in \chi\Delta$, then $\psi \neq 1$. Since $\hat{\nu}(\psi) = 0$ for $\psi \in F$, these observations and (3) show that $|\hat{\nu}(\psi)| \leqq \varepsilon$ for $\psi \in \chi\Delta \setminus \{\chi\}$. Hence we have

$$S_1 \leqq \varepsilon \|g\|_A < \frac{1}{4} \tag{9}$$

by (2). For $\psi \in X \setminus \chi\Delta$, we have $\chi\psi^{-1} \notin \Delta$ and so (3) and (4) lead to

$$S_2 \leqq \|\nu\| \sum_{\chi \in X\setminus\Delta} |\hat{g}(\chi)| < \frac{1}{4}. \tag{10}$$

From (8), (9) and (10) we obtain (7) at once. □

8.7 COROLLARY. <u>If</u> G <u>is</u> 0-<u>dimensional and if</u> E <u>is a symmetric Sidon set in</u> X, <u>then</u> E <u>is a full FZ-set if and only if</u> E <u>tends to infinity</u>.

<u>Proof</u>. See 8.6 and 8.5(c). □

Theorem 8.6 is true with the hypothesis "tends to infinity" weakened to "is almost X_0-subtransversal for all finite subgroups X_0 of X". This is the thrust of Theorem 8.16, for which several lemmas are needed. The first two give some technical properties of Sidon sets.

8.8 LEMMA. <u>If</u> E <u>is a Sidon set in</u> X, <u>then there is</u> $n \in \mathbb{Z}^+$ <u>such that</u>: <u>Given a finite set</u> $\Delta \subset X$, <u>there is a finite set</u> $F \subset E$ <u>satisfying</u>

$$|\chi\Delta \cap (E \setminus F)| \leq n \quad \underline{\text{for all}} \quad \chi \in X. \tag{1}$$

<u>Proof</u>. By Theorem 1.4, there is $n \in \mathbb{Z}^+$ such that

$$\sup\{\min(|A|, |B|) : AB \subset E\} \leq n. \tag{2}$$

Assume that (1) fails for this n. Then there is a finite set Δ in X such that to each finite set $F \subset E$ there corresponds $\chi_F \in X$ for which

$$|\chi_F\Delta \cap (E \setminus F)| > n. \tag{3}$$

We next show that there is a sequence $\{\chi_i\}_{i=1}^{\infty}$ of distinct characters in X such that

$$|\chi_i\Delta \cap E| > n \quad \text{for each} \quad i \in \mathbb{Z}^+. \tag{4}$$

First, let $F_1 = \emptyset$ and use (3) to obtain $\chi_1 \in X$ satisfying $|\chi_1\Delta \cap E| > n$. Suppose that distinct characters $\chi_1, \ldots, \chi_j$ have been chosen in X together with finite subsets $F_1 \subset F_2 \subset \cdots \subset F_j$ of E such that

$$|\chi_i\Delta \cap (E \setminus F_i)| > n \quad \text{for} \quad i = 1, 2, \ldots, j \tag{5}$$

and

$$\{\chi_i\Delta \cap (E \setminus F_i)\}_{i=1}^{j} \quad \text{is a pairwise disjoint family.} \tag{6}$$

Define $F_{j+1} = F_j \cup (\cup_{i=1}^{j}(\chi_i\Delta \cap E))$; hence $F_{j+1} \supset F_j$ and F_{j+1} is finite. Now use (3) to obtain $\chi_{j+1} \in X$ satisfying (5) for

$i = j+1$. For $i = 1,2,\ldots,j$, we have

$$[\chi_i \Delta \cap (E \setminus F_i)] \cap [\chi_{j+1} \Delta \cap (E \setminus F_{j+1})] \subset F_{j+1} \cap (E \setminus F_{j+1}) = \emptyset,$$

and so (6) holds for $j+1$. Moreover, χ_{j+1} is distinct from $\chi_1, \ldots, \chi_j$, for if $\chi_{j+1} = \chi_i$ for $i \leqq j$, then

$$[\chi_i \Delta \cap (E \setminus F_i)] \cap [\chi_{j+1} \Delta \cap (E \setminus F_{j+1})] = \chi_{j+1} \Delta \cap (E \setminus F_{j+1}) \neq \emptyset.$$

This completes our inductive construction of a sequence $\{\chi_i\}_{i=1}^{\infty}$ of distinct characters satisfying (4).

For each $i \in \mathbb{Z}^+$, we use (4) to select a set $B_i \subset \chi_i \Delta \cap E$ where $|B_i| = n+1$. Since Δ is finite, the sequence $\{\chi_i^{-1} B_i\}_{i=1}^{\infty}$ of $(n+1)$-element subsets of Δ must have finite range. Hence Δ contains an $(n+1)$-element subset B such that $\chi_i^{-1} B_i = B$ for an infinite set N of values of i. If $A = \{\chi_i\}_{i \in N}$, then $|A| = \aleph_0$, $|B| = n+1$ and

$$AB = \bigcup_{i \in N} \chi_i B = \bigcup_{i \in N} B_i \subset E,$$

contrary to (2). □

The next lemma is also due to Déchamps-Gondim. For Stechkin sets in $\mathbb{Z}$, it was proved by Gaposhkin [1967b; Lemma 4].

8.9 LEMMA. <u>Let</u> E <u>be a symmetric Sidon set in</u> X. <u>There is an integer</u> $m \in \mathbb{Z}^+$ <u>such that given a finite subset</u> Δ <u>of</u> X, E <u>can be written as a disjoint union</u> $E = F \cup (\bigcup_{i \in I} F_i)$ <u>where</u>

$$F \text{ and all } F_i \text{ are finite and symmetric,} \tag{1}$$

$$|F_i| \leqq m \text{ for all } i \in I, \tag{2}$$

<u>and</u>

$$F_i F_j \cap \Delta = \emptyset \text{ for } i \neq j, \; i,j \in I. \tag{3}$$

<u>Proof</u>. We may suppose that Δ is symmetric and $1 \in \Delta$. Let n be as in Lemma 8.8. By Lemma 8.8, there is a finite subset F of E such that

$$|\chi\Delta^{2n} \cap (E \setminus F)| \leqq n \quad \text{for all} \quad \chi \in X. \tag{4}$$

Clearly we may suppose that F is symmetric.

Let us call a sequence $\chi_1, \chi_2, \ldots, \chi_s$ in X a Δ-<u>chain</u> provided $s = 1$ or

$$\chi_{j+1} \in \chi_j\Delta \cup \chi_j^{-1}\Delta \quad \text{for} \quad j = 1,2,\ldots,s-1. \tag{5}$$

For $\chi, \psi \in E \setminus F$, we define $\chi \sim \psi$ if some Δ-chain in $E \setminus F$ contains them both. It is easy to see that $\sim$ is an equivalence relation on $E \setminus F$. We write $E \setminus F = \bigcup_{i \in I} F_i$ where the F_i are the equivalence classes defined by this equivalence relation. The sets F_i are evidently symmetric and property (3) is an immediate consequence of the definition of $\sim$. Thus it only remains to show (2) holds for suitable m that depends only on E.

We will prove (2) for $m = 2n$ by showing that $\sim$-equivalence classes in $E \setminus F$ have at most $2n$ elements. Consider χ, ψ in $E \setminus F$ where $\chi \sim \psi$. We show that

$$\text{for some} \quad \delta \in \{-1,1\}, \quad \psi^\delta \in \chi\Delta^{2n}. \tag{6}$$

We may suppose that $\chi \neq \psi$; hence there is a Δ-chain $\chi_1, \ldots, \chi_s$ in $E \setminus F$ such that $\chi_1 = \chi$, $\chi_s = \psi$, and $1 < s$. If the χ_j's are not distinct, then there are integers $1 \leqq i < k \leqq s$ such that $\chi_i = \chi_k$. If we remove $\chi_i, \ldots, \chi_{k-1}$ from the sequence $\chi_1, \chi_2, \ldots, \chi_s$, we obtain a shorter sequence having all the properties of the original sequence. Continuing in this fashion, we end up with a Δ-chain consisting of distinct elements, and so we may suppose that $\chi_1, \chi_2, \ldots, \chi_s$ are distinct. A simple induction on j shows that for $j = 1,2,\ldots,s$, there exist δ_j in $\{-1,1\}$ such that $\chi_j^{\delta_j}$ belongs to $\chi_1\Delta^{j-1}$; here $\Delta^0 = \{1\}$. The $\chi_j^{\delta_j}$ need not be distinct, but it is easy to see that no three of them can be equal. Hence $s \leqq 2n$, because otherwise the set

$$\{\chi_j^{\delta_j} : j = 1,2,\ldots,2n+1\}$$

has at least $n+1$ elements and is a subset of $\chi_1\Delta^{2n} \cap (E \setminus F)$, contrary to (4). Since $\psi^{\delta_s} = \chi_s^{\delta_s} \in \chi_1\Delta^{s-1} = \chi\Delta^{s-1} \subset \chi\Delta^{2n}$, (6) holds.

Assume that some $\sim$-equivalence class in $E \setminus F$ has $2n+1$ elements, say $\chi, \psi_1, \psi_2, \ldots, \psi_{2n}$. By (6), there are $\delta_1, \delta_2, \ldots, \delta_{2n}$ in $\{-1,1\}$ such that $\psi_j^{\delta_j} \in \chi\Delta^{2n}$. Then the set

$$\{\chi, \psi_1^{\delta_1}, \psi_2^{\delta_2}, \ldots, \psi_{2n}^{\delta_{2n}}\}$$

has at least $n+1$ elements and is a subset of $\chi\Delta^{2n} \cap (E \setminus F)$, again contrary to (4). □

8.10 COROLLARY. Lemma 8.9 remains valid if the word "symmetric" is removed twice. In addition, we can arrange for

$$F_i F_j^{-1} \cap \Delta = \emptyset \quad \text{for} \quad i \neq j, \quad i,j \in I. \tag{3'}$$

Proof. Apply Lemma 8.9 to $E \cup E^{-1}$. □

It is well known that if a trigonometric polynomial on a compact connected abelian group G vanishes on a set of positive measure, then it is identically zero. The next lemma is a generalization of this fact, since $\{1\}$ is the only finite subgroup of the character group X of a compact connected abelian group G. For $f \in \mathrm{Trig}(G)$, we write length(f) for the cardinal number of $\mathrm{Supp}(\hat{f}) = \{\chi \in X : \hat{f}(\chi) \neq 0\}$.

8.11 LEMMA. Let m be a positive integer, and let K be a λ-measurable subset of G with $\lambda(K) > 0$. There is a finite subgroup X_o of X such that

(i) if $f \in \mathrm{Trig}(G)$, $f\xi_K = 0$ λ-almost everywhere, length(f) $\leqq m$, and $\mathrm{Supp}(\hat{f})$ is X_o-subtransversal, then $f \equiv 0$.

Proof. Since the convolution

$$x \to \xi_K * \xi_{-K}(x) = \lambda(K \cap (K - x))$$

is a continuous function, so is the function

$$x \to \lambda(K) - \lambda(K \cap (K - x)) = \lambda(K \setminus (K - x)).$$

Consequently, there is a neighborhood W of 0 such that

$$\lambda(K \setminus (K - w)) < 2^{-m+1}\lambda(K) \quad \text{for } w \in W. \tag{1}$$

Let V be a symmetric neighborhood of 0 such that $(m - 1)V \subset W$, let G_o be the (open-and-closed) subgroup of G generated by V, and let X_o denote the annihilator in X of G_o. Note that G/G_o and its character group X_o are finite; see [HR; 23.25].

For this proof, we will call K itself a set of type 0 and a set of type $n + 1$ is any set of the form $K_n \cap (K_n - v)$ where K_n is of type n and $v \in V$. The present lemma follows from the assertion P_n below with $n = m$:

assertion P_n: If $f \in \mathrm{Trig}(G)$, $f\xi_{K_{m-n}} = 0$ λ-a.e. for some set K_{m-n} of type $m - n$, length(f) $\leqq n$, and $\mathrm{Supp}(\hat{f})$ is X_o-subtransversal, then $f \equiv 0$.

First we show that

sets of type $m-1$ have positive λ-measure. (2)

An easy induction on type shows that if K_n is of type n, then K_n has the form $\cap_{j=1}^{2^n}(K-x_j)$ where all x_j are in nV. In particular, a set of type $m-1$ has the form $\cap_{j=1}^{2^{m-1}}(K-w_j)$ where $w_j \in W$. If this set were to have λ-measure zero, then

$$\lambda(K) = \lambda(K \setminus \bigcap_{j=1}^{2^{m-1}} (K-w_j)),$$

and since

$$K \setminus \bigcap_{j=1}^{2^{m-1}} (K-w_j) \subset \bigcup_{j=1}^{2^{m-1}} (K \setminus (K-w_j))$$

(1) would lead to

$$\lambda(K) \leqq \sum_{j=1}^{2^{m-1}} \lambda(K \setminus (K-w_j)) < \lambda(K),$$

a contradiction. Hence (2) holds. Assertion P_1 follows immediately from (2) since a trigonometric polynomial of length $\leqq 1$ must be a multiple of a character.

Now we assume assertion P_n to be true for some n, $1 \leqq n \leqq m-1$, and prove assertion P_{n+1}. Consider f in $\mathrm{Trig}(G)$ and a set K' of type $m-(n+1)$ such that $f\xi_{K'} = 0$ λ-a.e., $\mathrm{length}(f) \leqq n+1$, and $\mathrm{Supp}(\hat{f})$ is X_0-subtransversal. Thus K' contains a set A' such that $\lambda(K' \setminus A') = 0$ and $f(x) = 0$ for all $x \in A'$. Assume that $f \not\equiv 0$; then $\mathrm{length}(f) = n+1$ by P_n. We write $f = a\chi_0 + g$ where $\chi_0 \in \mathrm{Supp}(\hat{f})$, $g \in \mathrm{Trig}(G)$, and $\mathrm{length}(g) = n$. We may suppose that $a = 1$. Consider a fixed $v \in V$ and observe that $K' \cap (K'-v)$ is a set of type $m-n$ and that

$$\lambda([K' \cap (K' - v)] \setminus [A' \cap (A' - v)]) = 0. \tag{3}$$

If x is in $A' \cap (A' - v)$, then $x, x + v \in A'$ and so

$$\begin{aligned}
0 &= f(x + v) - \chi_0(v)f(x) \\
&= \chi_0(x + v) + g(x + v) - \chi_0(v)[\chi_0(x) + g(x)] \\
&= g(x + v) - \chi_0(v)g(x) \\
&= \sum_{\chi \in X} \hat{g}(\chi)\chi(x + v) - \sum_{\chi \in X} \chi_0(v)\hat{g}(\chi)\chi(x) \\
&= \sum_{\chi \in X} \hat{g}(\chi)[\chi(v) - \chi_0(v)]\chi(x).
\end{aligned}$$

In view of (3), assertion P_n applies to the trigonometric polynomial $f_v = \sum_{\chi \in X} \hat{g}(\chi)\,[\chi(v) - \chi_0(v)]\,\chi$ and so $f_v \equiv 0$. Hence $\hat{g}(\chi)[\chi(v) - \chi_0(v)] = 0$ for all $\chi \in X$ and $v \in V$. Therefore if $\chi_1 \in \mathrm{Supp}(\hat{g})$, then $\chi_1(v) - \chi_0(v) = 0$ for all $v \in V$. It follows that $\chi_1\chi_0^{-1}(v) = 1$ for $v \in V$, hence $\chi_1\chi_0^{-1}(x) = 1$ for $x \in G_0$, and hence $\chi_1\chi_0^{-1} \in X_0$. Since χ_0 and χ_1 are distinct elements of $\mathrm{Supp}(\hat{f})$, this contradicts the X_0-subtransversality of $\mathrm{Supp}(\hat{f})$. □

8.12 COROLLARY. *Let* E *be a subset of* X *that is* X_0*-subtransversal for all finite subgroups* X_0 *of* X. *If* f *is in* $\mathrm{Trig}_E(G)$ *and* $\lambda(\{x \in G : f(x) = 0\}) > 0$, *then* $f \equiv 0$.

Proof. Given f, we apply Lemma 8.11 to $m = \mathrm{length}(f)$ and $K = \{x \in G : f(x) = 0\}$. □

8.13 COROLLARY. *If* G *is connected and if* f *in* $\mathrm{Trig}(g)$ *satisfies* $\lambda(\{x \in G : f(x) = 0\}) > 0$, *then* $f \equiv 0$.

Proof. Corollary 8.12 applies to $E = X$. □

The proof of Lemma 8.11 is descended from an argument in Bonami [1970; page 398]. The proof can be simplified if K has nonvoid interior; see Ross [1972; Lemme 1] or [1973;

Lemma 2.3]. The extra generality in 8.11 - 8.13 is not needed in this chapter, but will be used in Corollary 9.9 and Theorem 9.11. The next lemma goes back to Zygmund [1948]; see also Gaposhkin [1967b; Lemma 3], Déchamps-Gondim [1972; Lemme 6.3], and Ross [1973; Lemma 2.4].

8.14 LEMMA. _Let_ m _be a positive integer, and let_ g _be a nonzero function in_ $L^1_+(G)$. _There is a finite subgroup_ X_0 _of_ X _and_ $\varepsilon > 0$ _such that_

(i) _if_ $f \in \mathrm{Trig}(G)$, $\mathrm{length}(f) \leqq m$, _and_ $\mathrm{Supp}(\hat{f})$ _is_ X_0-_subtransversal, then_

$$\int_G |f|^2 g\, d\lambda \geqq \varepsilon \|f\|_2^2. \tag{1}$$

Proof. Let $K = \{x \in G : g(x) > 0\}$ and select X_0 as in Lemma 8.11. If (i) fails, there is a sequence $\{f_j\}_{j=1}^{\infty}$ in $\mathrm{Trig}(G)$ such that $\sup_j \mathrm{length}(f_j) \leqq m$, each $\mathrm{Supp}(\hat{f}_j)$ is X_0-subtransversal,

$$\lim_j \int_G |f_j|^2 g\, d\lambda = 0, \tag{2}$$

and

$$\lim_j \|f_j\|_2 = 1. \tag{3}$$

By passing to a subsequence we may assume that for some k in $\{1,2,\ldots,m\}$,

$$\mathrm{length}(f_j) = k \quad \text{for all } j.$$

We may also assume that k is _minimal_: no such sequence exists for smaller k. Observe that $k > 1$, since otherwise each f_j has the form $a_j \chi_j$ where $\chi_j \in X$ and $\lim_j |a_j| = 1$ so that

$$\lim_j \int_G |f_j|^2 g\, d\lambda = \int_G g\, d\lambda > 0.$$

Each f_j has the form

$$f_j = \sum_{i=1}^{k} \hat{f}_j(\chi_{ij})\chi_{ij}.$$

By replacing each f_j by $\overline{\chi_{1j}}f_j$, we may suppose that $\chi_{1j} = 1$ for all j. By re-ordering $\{1,2,\ldots,k\}$ if necessary, we may suppose that there is an integer r such that $1 \leqq r \leqq k$ and $\{\chi_{ij} : j = 1,2,\ldots\}$ is a finite set if and only if $1 \leqq i \leqq r$. By passing to further subsequences of $\{f_j\}_{j=1}^{\infty}$, we can also arrange for the following:

for $1 \leqq i \leqq r$, all χ_{ij} are equal (to χ_i, say); (4)

for $r < i \leqq k$, all χ_{ij} are distinct; (5)

for $1 \leqq i \leqq k$, the limit $\lim_j \hat{f}_j(\chi_{ij})$ exists (and equals c_i, say). (6)

[(6) is possible because $\lim\sup_j |\hat{f}_j(\chi_{ij})| \leq 1$ for all i by (3).] Note that $\chi_1, \chi_2, \ldots, \chi_r$ are distinct and that $\{\chi_1, \chi_2, \ldots, \chi_r\}$ is X_o-subtransversal. Let

$$f = \sum_{i=1}^{r} c_i \chi_i.$$

Case 1. If $r = k$, then $\lim_j f_j = f$ uniformly by (6) and (4), and hence (2) yields

$$\int_G |f|^2 g \, d\lambda = \lim_j \int_G |f_j|^2 g \, d\lambda = 0.$$

It follows that f vanishes λ-a.e. on K, and so by 8.11(i) f vanishes identically. On the other hand,

$$\|f\|_2 = \lim_j \|f_j\|_2 = 1$$

by (3), which is a contradiction.

Case 2. Suppose that $r < k$ and $f \equiv 0$. For each j, let

$$F_j = f_j - \sum_{i=1}^{r} \hat{f}_j(\chi_i)\chi_i = \sum_{i=r+1}^{k} \hat{f}_j(\chi_{ij})\chi_{ij}.$$

Since $\lim_j \hat{f}_j(\chi_i) = c_i = 0$ for $i \leq r$, we have

$$\lim_j (F_j - f_j) = 0 \text{ uniformly.}$$

It follows from (2) and (3) that $\lim_j \int_G |F_j|^2 g\, d\lambda = 0$ and $\lim_j \|F_j\|_2 = 1$. The existence of the sequence $\{F_j\}_{j=1}^{\infty}$ with these properties violates the minimality of k.

Case 3. Suppose that $r < k$ and f does not vanish identically. By 8.11(i), $f\xi_K(x) \neq 0$ for x in a set of positive λ-measure and so $\int_G |f|^2 g\, d\lambda > 0$. Let F_j be as in Case 2. By Hölder's inequality, we have

$$\left|\int_G \overline{f_j}\, f\, g\, d\lambda\right|^2 \leq \left(\int_G |f_j|^2 g\, d\lambda\right)\left(\int_G |f|^2 g\, d\lambda\right)$$

and so by (2), we have $\lim_j \int_G \overline{f_j}\, f\, g\, d\lambda = 0$. Since we also have $\lim_j (f_j - F_j) = f$ uniformly, we conclude that

$$\lim_j \int_G \overline{F_j}\, f\, g\, d\lambda = \lim_j \int_G \overline{f_j}\, f\, g\, d\lambda - \int_G \overline{f} f g\, d\lambda \neq 0. \tag{7}$$

We can also write

$$\int_G \overline{F_j}\, f\, g\, d\lambda = \sum_{i=r+1}^{k} \overline{\hat{f}_j(\chi_{ij})} \int_G \overline{\chi_{ij}}\, f\, g\, d\lambda$$

$$= \sum_{i=r+1}^{k} \overline{\hat{f}_j(\chi_{ij})}\, (fg)\hat{\ }(\chi_{ij}). \tag{8}$$

Since fg belongs to $L^1(G)$, the Riemann-Lebesgue lemma and (5) imply that $\lim_j (fg)\hat{\ }(\chi_{ij}) = 0$ for $i = r+1,\ldots,k$. Therefore

$$\lim_j \int_G \overline{F_j}\, f\, g\, d\lambda = 0$$

by (8) and (6), and this contradicts (7).

All three cases lead to contradictions and so (i) must hold. □

8.15 LEMMA. <u>Let</u> m <u>be a positive integer, and let</u> g <u>be a nonzero function in</u> $L^1_+(G)$. <u>There is a finite subgroup</u> X_0 <u>of</u> X <u>and</u> $\delta > 0$ <u>such that</u>

(i) <u>if</u> $\{\chi_1, \chi_2, \ldots, \chi_k\}$ <u>is an</u> X_0<u>-subtransversal subset of</u> X <u>with</u> $2 \leqq k \leqq m$, <u>then the determinant of the matrix</u> $A = [\hat{g}(\chi_i \chi_j^{-1})]_{i,j=1}^k$ <u>is greater than or equal to</u> δ.

<u>Proof</u>. Clearly we may assume $m \geqq 2$. Let X_0 and ε be as given by Lemma 8.14, and note that we may assume that $\varepsilon \leqq 1$. Let $\delta = \varepsilon^m$. Since $\hat{g}$ is a positive-definite function on X, the matrix A is positive-definite and its determinant is a product of eigenvalues of A. Hence it suffices to show that $\alpha \geqq \varepsilon$ for each eigenvalue α of A. Let $c = (c_1, c_2, \ldots, c_k)$ be an eigenvector for α: $Ac = \alpha c$. Lemma 8.14 applies to the trigonometric polynomial $f = \Sigma_{i=1}^k c_i \chi_i$ and so

$$\alpha \|f\|_2^2 = \alpha \sum_{i=1}^k |c_i|^2 = \langle \alpha c, c\rangle = \langle Ac, c\rangle = \sum_{i,j=1}^k \overline{c_i}\, c_j\, \hat{g}(\chi_i \chi_j^{-1})$$

$$= \int_G \sum_{i,j=1}^k \overline{c_i}\, c_j\, \overline{\chi_i}\, \chi_j\, g\, d\lambda = \int_G |f|^2 g\, d\lambda \geqq \varepsilon \|f\|_2^2 .$$

Thus $\alpha \geqq \varepsilon$ as desired. □

We are finally able to state and prove our main theorem. Some interesting corollaries follow the theorem.

8.16 THEOREM. A symmetric subset E of X is a full FZ-set if and only if

(i) E is a Sidon set,

and

(ii) E is almost X_0-subtransversal for all finite subgroups X_0 of X.

Proof. If E is a symmetric full FZ-set, then (i) holds by 7.4 and 2.3, and (ii) holds by Lemma 8.3(i).

Suppose now that (i) and (ii) hold. Let K be a compact subset of G with nonvoid interior. We will show that E is an FZ(K)-set. Let g be a nonnegative function in A(G) such that

$$\mathrm{Supp}(g) \subset -K \quad \text{and} \quad \hat{g}(1) = \int_G g \, d\lambda = 1; \tag{1}$$

the existence of g follows from the regularity of the algebra A(G) [R; 2.6.2] and the fact that -K has nonvoid interior. Let m be as given by Lemma 8.9. For this m and g, let X_0 and $\delta > 0$ be as in Lemma 8.15, and let $C = \max(1, m!/\delta)$. Let $\varepsilon > 0$ be so that $\varepsilon < 1$ and

$$4\varepsilon \|g\|_A < 1. \tag{2}$$

By Drury's Theorem 3.3, there is a constant $M > 0$ such that if ϕ_1 is hermitian on E and $\|\phi_1\|_\infty \leq C$, then there is $\nu \in M^+(G)$ satisfying:

$$\hat{\nu}|_E = \phi_1, \quad \|\nu\| \leq M, \quad \text{and} \quad |\hat{\nu}(\chi)| \leq \varepsilon \quad \text{for} \quad \chi \notin E \cup \{1\}. \tag{3}$$

[If $\varkappa$ is a Sidon constant for E, M can be taken to be $32\varkappa^4 C^2/\varepsilon$.]

Now let Δ be a finite symmetric subset of X containing 1 and satisfying

$$4M \sum_{\chi \in X\setminus\Delta} |\hat{g}(\chi)| < 1. \tag{4}$$

Let F_o be a finite symmetric subset of E such that $E\setminus F_o$ is X_o-subtransversal and $(E\setminus F_o)\cap\Delta = \emptyset$. For Δ as above, decompose $E\setminus F_o$ as in Lemma 8.9: $E\setminus F_o$ is a disjoint union $F\cup(\bigcup_{i\in I} F_i)$ where F and all F_i are finite and symmetric, $|F_i| \leqq m$ for all $i \in I$, and

$$F_iF_j \cap \Delta = \emptyset \quad \text{for} \quad i \neq j, \quad i,j \in I. \tag{5}$$

Let $E_o = F_o \cup F$ and note that $E\setminus E_o = \bigcup_{i\in I} F_i$ is X_o-subtransversal and disjoint from Δ.

Let $\phi : E \to \mathbb{T}$ be a hermitian function. By 7.2(vii), it suffices for us to find μ in $M^+(-K)$ such that

$$|u(\chi) - \phi(\chi)| \leqq \tfrac{1}{4} \quad \text{for all} \quad \chi \in E\setminus E_o. \tag{6}$$

We now define ϕ_1 on $E\setminus E_o = \bigcup_{i\in I} F_i$ so that $\|\phi_1\|_\infty \leqq C$ as follows. For $i \in I$ write $F_i = \{\chi_{i1},\ldots,\chi_{ik}\}$ where $1 \leqq k \leqq m$, and consider the linear system

$$\sum_{s=1}^{k} x_s\, \hat{g}(\chi_{ir}\chi_{is}^{-1}) = \phi(\chi_{ir}) \qquad (1 \leqq r \leqq k). \tag{7}$$

If F_i consists of a single element χ_i, then the solution to (7) is $x_1 = \phi(\chi_i)$ by (1) and we define $\phi_1(\chi_i) = \phi(\chi_i)$. In this case we have

$$\chi_i^2 = 1 \quad \text{and} \quad \phi_1(\chi_i) = \phi(\chi_i) \quad \text{is a real number.} \tag{8}$$

Otherwise, $2 \leqq k \leqq m$ and Lemma 8.15 asserts that the determinant of the system (7) is $\geqq \delta$. Hence its unique solution $(x_1,\ldots,x_k)$ satisfies $|x_s| \leqq m!/\delta$ for $1 \leqq s \leqq k$ as can be seen from Cramer's rule. We define $\phi_1(\chi_{is}) = x_s$ for

$1 \leq s \leq k$, so that

$$\sum_{s=1}^{k} \phi_1(\chi_{is})\hat{g}(\chi_{ir}\chi_{is}^{-1}) = \phi(\chi_{ir}) \qquad (1 \leq r \leq k). \tag{9}$$

Note first that $\|\phi_1\|_\infty \leq C$. Since F_i is symmetric, there is a permutation π of $\{1,2,\ldots,k\}$ such that $\chi_{i,\pi(s)} = \chi_{is}^{-1}$ for all $s \in \{1,2,\ldots,k\}$. Replacing all s and r in (9) by $\pi(s)$ and $\pi(r)$, respectively, we obtain

$$\sum_{s=1}^{k} \phi_1(\chi_{is}^{-1})\hat{g}(\chi_{ir}^{-1}\chi_{is}) = \phi(\chi_{ir}^{-1}).$$

Since $\hat{g}$ and ϕ are hermitian, we have

$$\sum_{s=1}^{k} \phi_1(\chi_{is}^{-1})\overline{\hat{g}(\chi_{ir}\chi_{is}^{-1})} = \overline{\phi(\chi_{ir})}$$

for $1 \leq r \leq k$, and so $x_s = \overline{\phi_1(\chi_{is}^{-1})}$ $(s = 1,2,\ldots,k)$ is a solution of (7). Since the solution of (7) is unique, we infer that

$$\phi_1(\chi) = \overline{\phi_1(\chi^{-1})} \quad \text{for all} \quad \chi \in F_i. \tag{10}$$

Define ϕ_1 to be zero on E_0. Then ϕ_1 is a hermitian function on E by (8) and (10). It follows that there is a ν in $M^+(G)$ satisfying (3).

Finally, let $\mu = g\nu$; clearly μ is in $M^+(-K)$ since $\mathrm{Supp}(g) \subset -K$ by (1). It remains for us to establish (6). Consider a fixed $\chi_0 \in F_i \subset E \setminus E_0$. As before, write $F_i = \{\chi_{i1}, \chi_{i2}, \ldots, \chi_{ik}\}$, $1 \leq k \leq m$, so that χ_0 equals χ_{ir} for some $r \in \{1,2,\ldots,k\}$. By (9) and (3), we have

$$\begin{aligned} \phi(\chi_0) = \phi(\chi_{ir}) &= \sum_{s=1}^{k} \phi_1(\chi_{is})\hat{g}(\chi_{ir}\chi_{is}^{-1}) \\ &= \sum_{\chi \in F_i} \phi_1(\chi)\hat{g}(\chi_0\chi^{-1}) = \sum_{\chi \in F_i} \hat{\nu}(\chi)\hat{g}(\chi_0\chi^{-1}), \end{aligned}$$

whereas

$$\hat{\mu}(\chi_0) = \int_G \chi_0^{-1} g\, d\nu = \sum_{\chi\in X} \hat{g}(\chi) \int_G \chi_0^{-1} \chi\, d\nu$$

$$\sum_{\chi\in X} \hat{g}(\chi)\hat{\nu}(\chi_0\chi^{-1}) = \sum_{\chi\in X} \hat{g}(\chi_0\chi^{-1})\hat{\nu}(\chi).$$

Therefore we have

$$|\hat{\mu}(\chi_0) - \phi(\chi_0)| \leqq \sum_{\chi\in X\setminus F_i} |\hat{g}(\chi_0\chi^{-1})|\cdot|\hat{\nu}(\chi)| \equiv S_1 + S_2, \qquad (11)$$

where S_1 is the sum over $\chi_0\Delta\setminus F_i$ and S_2 is the sum over $X\setminus(F_i \cup \chi_0\Delta)$.

Next consider $\chi \in \chi_0\Delta\setminus F_i$. Then $\chi \notin E\setminus E_0$, since otherwise χ would belong to F_j for some $j \neq i$ and then $\chi\chi_0^{-1} \in F_jF_i^{-1} \cap \Delta$, contrary to (5). Also $\chi \neq 1$, since otherwise χ_0^{-1} would belong to $(E\setminus E_0)\cap\Delta = \emptyset$. If $\chi \in E_0$, then we have $\hat{\nu}(\chi) = \phi_1(\chi) = 0$; otherwise $\chi \notin E \cup \{1\}$, and so $|\hat{\nu}(\chi)| \leqq \varepsilon$ by (3). Hence using (2), we obtain

$$S_1 \leqq \sum_{\chi\in\chi_0\Delta\setminus F_i} |\hat{g}(\chi_0\chi^{-1})|\cdot\varepsilon \leqq \varepsilon\|g\|_A \leqq \tfrac{1}{4}.$$

Finally we use $\|\nu\| \leqq M$ (from (3)) and (4) to write

$$S_2 \leqq \sum_{\chi\in X\setminus\chi_0\Delta} |\hat{g}(\chi_0\chi^{-1})|\cdot|\hat{\nu}(\chi)| = \sum_{\psi\in X\setminus\Delta} |\hat{g}(\psi)|\cdot|\hat{\nu}(\chi_0\psi^{-1})|$$

$$\leqq \|\nu\| \sum_{\psi\in X\setminus\Delta} |\hat{g}(\psi)| \leqq \tfrac{1}{4}.$$

Thus (11) shows that $|\hat{\mu}(\chi_0) - \phi(\chi_0)| \leqq \frac{1}{2}$ and so (6) holds. □

8.17 COROLLARY. <u>Let</u> G <u>be a compact connected abelian group with character group</u> X. <u>A symmetric subset</u> E <u>of</u> X <u>is a full FZ-set if and only if it is a Sidon set</u>.

<u>Proof</u>. If G is connected, 8.16(ii) trivially holds for all subsets E of X. □

8.18 COROLLARY. _Let_ G _be any compact abelian group, and let_ E _be a subset of_ X. _The following are equivalent:_

(i) E _is associated with every compact subset of_ G _with nonvoid interior;_

(ii) E _is a Sidon set that is almost_ X_0_-subtransversal for all finite subgroups_ X_0 _of_ X.

Proof. If (i) holds, then E is associated with G itself and so E is a Sidon set. This observation and Lemma 8.3(i) show that (i) implies (ii). Suppose that (ii) holds. In order to apply Theorem 8.16 we resort to the device employed in the proof of Corollary 3.4. Thus we let $G_0 = G \times \mathbb{T}$, but we write Y for its character group $X \times \mathbb{Z}$. Let $E_0 = E \times \{1\}$. We showed that $E_0 \cup E_0^{-1}$ is a Sidon set in the proof of 3.4. Next we show that $E_0 \cup E_0^{-1}$ is almost Y_0-subtransversal for finite subgroups Y_0 of Y. Note that $Y_0 = X_0 \times \{0\}$ for some finite subgroup X_0 of X. So there is a finite subset F of E such that $E \setminus F$ is X_0-subtransversal. Let $F_0 = F \times \{1\}$. It is now easy to check that the set $E_0 \cup E_0^{-1} \setminus (F_0 \cup F_0^{-1})$ is Y_0-subtransversal.

Theorem 8.16 implies that $E_0 \cup E_0^{-1}$ is a full FZ-set. Hence, as noted in the last paragraph of the proof of 8.3, $E_0 \cup E_0^{-1}$ is associated with every compact set in G_0 with nonvoid interior. In particular, E_0 is associated with every compact set in G_0 with nonvoid interior. Let K be a compact set in G with nonvoid interior. Then there is a finite subset F_0 of E_0 and a constant $\varkappa > 0$ such that

$$\|f\|_A \leq \varkappa \|f \xi_{K \times \mathbb{T}}\|_u \quad \text{for } f \text{ in } \mathrm{Trig}_{E_0 \setminus F_0}(G_0). \tag{1}$$

Let F be a finite subset of E such that $F >< \{1\} \supset F_o$ and consider g in $\mathrm{Trig}_{E\setminus F}(G)$. Then $g = \sum_{k=1}^{n} c_k\chi_k$ where $\chi_k \in E \setminus F$ and so $f = \sum_{k=1}^{n} c_k(\chi_k, 1)$ belongs to $\mathrm{Trig}_{E_o\setminus F_o}(G_o)$. Then (1) implies that $\|g\|_A \leqq \varkappa\|g\xi_K\|_u$ since $f(x,z) = g(x)z$ for $(x,z) \in G_o$. This proves that $E \setminus F$ is strictly associated with K, and so (i) holds. □

8.19 COROLLARY [DÉCHAMPS-GONDIM]. <u>Let</u> G <u>be a compact connected abelian group. A set</u> E <u>in</u> X <u>is a Sidon set if and only if it is associated with every compact subset of</u> G <u>with nonvoid interior.</u>

Corollary 8.19 is the first theorem of Déchamps-Gondim mentioned on page 109. We now turn our attention to proving a generalization of her second theorem. The next lemma is a simple consequence of the open mapping theorem.

8.20 LEMMA. <u>Let</u> E <u>be a Sidon set in</u> X. <u>For a closed set</u> $K \subset G$, <u>let</u> $U_E(K)$ <u>denote the uniform closure in</u> $C(K)$ <u>of</u> $\{f|_K : f \in \mathrm{Trig}_E(G)\}$. <u>For each</u> $f \in C_E(G)$, <u>let</u> $\rho(f) = f|_K$. <u>Then</u> ρ <u>is a bounded linear map of</u> $C_E(G)$ <u>into</u> $U_E(K)$. <u>Also</u> ρ <u>is one-to-one and onto if and only if</u> K <u>and</u> E <u>are strictly associated.</u>

<u>Proof</u>. Let $\{h_\alpha\} \subset \mathrm{Trig}^+(G)$ be an approximate unit for $L^1(G)$ such that $\|h_\alpha\|_1 = 1$ for all α. If f belongs to $C_E(G)$, then $\lim_\alpha\|f*h_\alpha - f\|_u = 0$ and each $f*h_\alpha$ is in $\mathrm{Trig}_E(G)$. Since $\rho(f) = f|_K$ is the uniform limit of the net $\{f*h_\alpha|_K\}$, it is clear that $\rho(f) \in U_E(K)$. Clearly ρ is linear and bounded; in fact, $\|\rho(f)\|_u \leqq \|f\|_u$ for $f \in C_E(G)$.

Suppose that ρ is one-to-one and onto. By the open mapping theorem, there exists $\kappa > 0$ so that $\|f\|_u \leqq \kappa\|f\,\xi_K\|_u$ for all $f \in C_E(G)$. Since E is a Sidon set, 1.3(vii) provides us with a constant $\kappa_0 > 0$ such that $\|f\|_A \leqq \kappa_0\|f\|_u$ for all $f \in \mathrm{Trig}_E(G)$. Hence we have $\|f\|_A \leqq \kappa_0\kappa\|f\,\xi_K\|_u$ for all f in $\mathrm{Trig}_E(G)$, i.e. E and K are strictly associated.

Now suppose that E and K are strictly associated. Then $\|f\|_u \leqq \|f\|_A \leqq \kappa\|f\,\xi_K\|_u$ for all $f \in \mathrm{Trig}_E(G)$ and some constant $\kappa > 0$. The argument at the beginning of the proof shows that $\mathrm{Trig}_E(G)$ is uniformly dense in $C_E(G)$. Hence we have $\|f\|_u \leqq \kappa\|f\,\xi_K\|_u$ for all $f \in C_E(G)$. It follows that ρ is a one-to-one bicontinuous linear map. Since $\rho[C_E(G)]$ is closed and dense in $U_E(K)$, ρ is also an onto map. □

8.21 LEMMA. *Let $E \subset X$ be a Sidon set and suppose that χ_0 in $X \setminus E$ has the property that $E \cup \{\chi_0\}$ is X_0-subtransversal for all finite subgroups X_0 of X. Let K be a closed subset of G strictly associated with E, and let W be a closed set in G containing 0 and having positive λ-measure. Then $K + W$ is strictly associated with $E \cup \{\chi_0\}$.*

Proof. Let $K_0 = K + W$ and $E_0 = E \cup \{\chi_0\}$. By Lemma 8.20 and our hypothesis, the restriction map $\rho : C_E(G) \to U_E(K)$ is one-to-one and onto. The same lemma shows that it suffices to prove that the restriction map $\rho_0 : C_{E_0}(G) \to U_{E_0}(K_0)$ is one-to-one and onto. Corollary 8.12 shows that

$$f \in \mathrm{Trig}_{E_0}(G) \text{ and } f|_W = 0 \text{ imply } f \equiv 0. \tag{1}$$

Since ρ is one-to-one, we also have

$$f \in C_E(G) \text{ and } f|_K = 0 \text{ imply } f \equiv 0. \tag{2}$$

To see that ρ_o is one-to-one, consider $f \in C_{E_o}(G)$ such that $f|_{K_o} = 0$. Then we have $f(x+w) - \chi_o(w)f(x) = 0$ for all $x \in K$ and $w \in W$. For each $w \in W$, we define

$$g_w(x) = f(x+w) - \chi_o(w)f(x) \quad \text{for all} \quad x \in G.$$

A simple computation shows that

$$\hat{g}_w(\chi) = [\chi(w) - \chi_o(w)]\hat{f}(\chi) \quad \text{for all} \quad \chi \in X.$$

Hence each g_w belongs to $C_E(G)$. Also, $g_w|_K = 0$ and so $g_w \equiv 0$ by (2). Assume that $\hat{f}(\chi) \neq 0$ for some $\chi \in E$. Since $\hat{g}_w \equiv 0$ for all $w \in W$, we must have $\chi(w) - \chi_o(w) = 0$ for all $w \in W$. Therefore $\chi - \chi_o \equiv 0$ by (1) and so $\chi_o \in E$, a contradiction. Thus $f = c\chi_o$ for some constant c. Since $f|_{K_o} = 0$, we must have $c = 0$ and $f \equiv 0$. That is, ρ_o is one-to-one.

To see that ρ_o is an onto map, consider $g \in U_{E_o}(K_o)$. Then there exists a sequence $\{f_n\}$ in $\mathrm{Trig}_E(G)$ and a sequence $\{a_n\}$ of complex numbers such that

$$g = \lim_n (f_n + a_n\chi_o)|_{K_o} \quad \text{uniformly}. \tag{3}$$

We first claim that

$$\{a_n\} \text{ is a bounded sequence.} \tag{4}$$

Otherwise we may assume without loss of generality that $\lim_n |a_n| = \infty$. Then by (3) we have

$$\lim_n (a_n^{-1} f_n + \chi_o)|_{K_o} = 0 \quad \text{uniformly}. \tag{5}$$

Hence $\{a_n^{-1} f_n\}$ is a uniformly Cauchy sequence of functions on K_o, and hence on K. Since K and E are strictly associated, we have

$$\|f\|_u \leq \|f\|_A \leq \varkappa \|f \xi_K\|_u \quad \text{for all} \quad f \in \mathrm{Trig}_E(G) \tag{6}$$

and for some constant $\varkappa > 0$. Hence $\{a_n^{-1} f_n\}$ is uniformly Cauchy on G and there exists $h \in C(G)$ such that $\lim_n \|a_n^{-1} f_n - h\|_u = 0$. Clearly h belongs to $C_E(G)$. By (5) we have $h(y) = -\chi_0(y)$ for $y \in K_0$, i.e. $\rho_0(h) = \rho_0(-\chi_0)$. Since ρ_0 is one-to-one, we conclude that $\chi_0 = -h \in C_E(G)$, a contradiction. Thus (4) holds.

We may now assume without loss of generality that $\lim_n a_n = a$ exists and is finite. Then $\{f_n\}$ is uniformly Cauchy on K_0 and K by (3), and (6) shows that $\{f_n\}$ is uniformly Cauchy on G. Thus there exists $f \in C_E(G)$ such that $\lim_n \|f - f_n\|_u = 0$. Clearly $f + a\chi_0$ belongs to $C_{E_0}(G)$ and

$$\rho_0(f + a\chi_0) = (f + a\chi_0)|_{K_0} = g$$

by (3). This proves that ρ_0 is an onto map and completes the proof. □

8.22 THEOREM. *The following are equivalent for a subset* E *of* X.

(i) E *is strictly associated with every compact subset of* G *with nonvoid interior*.

(ii) E *is a Sidon set that is* X_0*-subtransversal for all finite subgroups* X_0 *of* X.

Proof. If (i) holds, then E is strictly associated with G, and so E is a Sidon set. This combines with Lemma 8.3(ii) to show that (ii) holds.

Suppose that (ii) holds and consider a compact subset K of G with nonvoid interior $\text{int}(K)$. Let x_0 be in $\text{int}(K)$. Let V be a closed neighborhood of 0 such that $x_0 + V + V \subset K$. By Corollary 8.18, $x_0 + V$ is associated with E.

So there exist $\chi_1, \chi_2, \ldots, \chi_m$ in E such that $x_0 + V$ is strictly associated with $E \setminus \{\chi_1, \chi_2, \ldots, \chi_m\}$. Let W be a closed neighborhood of 0 such that $mW \subset V$. By Lemma 8.21, $x_0 + V + W$ is strictly associated with $E \setminus \{\chi_2, \ldots, \chi_m\}$ and, by a simple induction, $x_0 + V + mW$ is strictly associated with E. Since $x_0 + V + mW \subset K$, we see that K and E are strictly associated. □

8.23 COROLLARY [DÉCHAMPS-GONDIM]. _Let_ G _be a compact connected abelian group. A set_ E _in_ X _is a Sidon set if and only if it is strictly associated with every compact subset of_ G _with nonvoid interior._

It is natural to inquire whether any of the results 8.18, 8.19, 8.22 or 8.23 hold if "every compact subset of G with nonvoid interior" is replaced by "every compact subset of G with positive Haar measure". Déchamps-Gondim [1972; Remarque 6.2] shows that the answer is always negative; see 8.24 below. However, improvements are possible if the notion of "associated" is replaced by "2-associated"; see 9.11 and 9.12.

*8.24 PROPOSITION. _Let_ E _be an infinite Sidon set in_ X _and_ $\varepsilon > 0$. _Then_ G _contains a closed set_ K _such that_ $\lambda(K) > 1 - \varepsilon$ _and such that_ E _is not associated with_ K.

Proof. We may suppose that E is countable and enumerate E as $\{\chi_n\}_{n=1}^{\infty}$. Let $\{a_n\}_{n=1}^{\infty}$ be any sequence of complex numbers such that $\sum_{n=1}^{\infty} |a_n|^2 < \infty$ and $\sum_{n=1}^{\infty} |a_n| = \infty$, and let $f = \sum_{n=1}^{\infty} a_n \chi_n \in L^2(G)$. For each $m \in \mathbb{Z}^+$, let $f_m = \sum_{n=1}^{m} a_n \chi_n$. Parseval's identity shows that $\lim_{m \to \infty} \|f - f_m\|_2 = 0$, and so $f_m \to f$ in measure. By Riesz's theorem, a subsequence f_{m_k}

converges λ-a.e. to f. By Egorov's theorem, there is a closed set $K \subset G$ such that $f_{m_k} \to f$ uniformly on K and $\lambda(K) > 1 - \varepsilon$. Assume that E is associated with K. For some N, $E_N = \{\chi_n\}_{n=N}^{\infty}$ is strictly associated with K. Thus there is $\varkappa > 0$ so that

$$\|h\|_A \leqq \varkappa \|h\, \xi_K\|_u \quad \text{for } h \text{ in } \mathrm{Trig}_{E_N}(G).$$

This inequality applies to $f_{m_k} - f_{m_j}$ whenever $m_k \geqq N$ and $m_j \geqq N$, and so $\{f_{m_k}\}_{k=1}^{\infty}$ is a Cauchy sequence in $A(G)$. Clearly the limit must be f and so f belongs to $A(G)$. Thus $\sum_{n=1}^{\infty}|a_n| < \infty$, contrary to our earlier choice. □

*8.25 REMARKS. (a) Let G be connected and let E be a Sidon set in X. For every compact subset K of G with nonvoid interior, Corollary 8.23 shows that

$$\|f\|_u \leqq \varkappa \|f\, \xi_K\|_u \quad \text{for all } f \in \mathrm{Trig}_E(G) \tag{1}$$

and some constant $\varkappa$ depending on K. Note that we have written $\|f\|_u$ instead of $\|f\|_A$ on the left side of (1). Déchamps-Gondim [1972; Remarque 6.1] points out that non-Sidon sets in $\mathbb{Z}$ can have this property. Blei [1973] proves that every non-Sidon set contains a non-Sidon set E with the following striking property: To each $\varepsilon > 0$ and compact set $K \subset G$ with nonvoid interior, there corresponds a finite subset F of E such that

$$\|f\|_u \leqq (1+\varepsilon)\|f\, \xi_K\|_u \quad \text{for all } f \in \mathrm{Trig}_{E\setminus F}(G). \tag{2}$$

(b) The following theorem is stated by Déchamps-Gondim [1970a; Théorème 2]. Let E be a finite union of Rider sets and assume that $E \cup E^{-1}$ tends to infinity. Then there is a constant $\varkappa > 0$ depending only on E such that given a compact set $K \subset G$ with nonvoid interior, there is a finite subset F of E for which

$$\|f\|_A \leqq \varkappa \|f \xi_K\|_u \quad \text{for all} \quad f \in \mathrm{Trig}_{E \setminus F}(G). \tag{3}$$

(c) Recall that if E_0 is a dissociate set, then $E = E_0 \cup E_0^{-1}$ tends to infinity; see 8.5(d). Hence, as noted in 8.5(b), E satisfies 8.16(ii). Theorem 2.7 shows that E is a Sidon set. Therefore E is a full FZ-set by Theorem 8.16. It follows that infinite full FZ-sets exist in every infinite group X; see 2.8. This fact was observed by Ross [1973]. The examples in (d) and (e) below are also taken from Ross [1973].

(d) Corollary 2.9 shows that every infinite subset of X contains infinite Sidon sets. This result does not extend to full FZ-sets. Select G so that X has a character ψ of order 3 and an infinite sequence $\{\chi_k\}_{k=1}^{\infty}$ of characters of order 2; a specific example appears in part (f). Let $\Delta = \{\psi\chi_k : k = 1,2,\ldots\}$, so that $\Delta^{-1} = \{\psi^2\chi_k : k = 1,2,\ldots\}$, and let $E = \Delta \cup \Delta^{-1}$. No infinite symmetric subset of E is almost X_0-subtransversal for the finite subgroup $X_0 = \{1, \psi, \psi^2\}$. Hence by Theorem 8.16, E contains no infinite full FZ-subsets.

(e) In contrast to Drury's Theorem 3.5, the union of two full FZ-sets need not be a full FZ-set. In fact, suppose that X has the property that $Y = \{\chi \in X : \chi^2 = 1\}$ is infinite. Since Y is a direct sum of two-element groups [HR; A.25], there exist a two-element subgroup $X_o = \{1, \chi_o\}$ and an infinite subgroup Y_o of Y such that $X_o \cap Y_o = \{1\}$. As shown in Theorem 2.8, Case 2, Y_o contains an infinite dissociate set E_1. It is easy to see that $E_2 = \chi_o E_1$ is also dissociate. Since E_1 and E_2 are symmetric, part (c) shows that these sets are full FZ-sets. Their union $X_o E_1$ is not almost X_o-subtransversal and so it is not a full FZ-set by Theorem 8.16.

(f) Here is an example of a Rider set that does not tend to infinity. Let G be the infinite product

$$\mathbb{Z}(3) \times \mathbb{Z}(2) \times \mathbb{Z}(2) \times \mathbb{Z}(2) \times \cdots$$

and X its character group. For each n, let π_n denote the n-th projection, and define

$$\Delta = \{\pi_1 \pi_n : n = 2,3,4,\ldots\} \quad \text{and} \quad E = \Delta \cup \Delta^{-1}.$$

A nonvoid asymmetric subset S of E has the form

$$\{\pi_1 \pi_{n_k} : k = 1,2,\ldots,t\} \cup \{\pi_1^2 \pi_{n_k} : k = t+1,\ldots,s\}$$

where $0 \leqq t \leqq s$ and $\{n_k\}_{k=1}^{s}$ is a sequence of distinct integers $\geqq 2$. If the product of characters in S is 1, then also $\prod_{k=1}^{s} \pi_{n_k} = 1$, an impossibility. Thus we have $R_s(E,1) = 0$ for all $s \geqq 1$, so that E is a Rider set. Since $(\pi_1 \pi_n)^2 = \pi_1^2 \neq 1$ for all $n \geqq 2$, 8.5(e) shows that E cannot tend to infinity.

Chapter 9

MISCELLANEOUS TOPICS

The bulk of this chapter is devoted to sets E in X that are 2-associated or strictly 2-associated to sets in G. These are notions analogous to sets associated or strictly associated to subsets of G; compare Definitions 9.3 and 8.1. The connection with Sidon sets is established in Theorem 9.11 and its corollaries. These results will allow us to prove that if G is connected, if E is a Sidon set in X, and if an E-spectral trigonometric series $\sum_{\chi \in E} \phi(\chi)\chi$ converges on a set of positive measure, then ϕ is in $\ell^2(E)$. If this series converges on a set with nonvoid interior, then ϕ belongs to $\ell^1(E)$. See Corollary 9.16.

We first prove an interesting corollary to Lemma 8.9 due to Déchamps-Gondim [1972; Cor. to Lemme 6.2]. The reader may wish to skip the proof, since Theorem 9.1 will not be used in any subsequent results. Déchamps-Gondim kindly provided the details of the proof in private correspondence.

9.1 THEOREM. Every countable Sidon set E in X is a finite union of Sidon sets tending to infinity. [For the definition, see 8.4.]

Proof. We may suppose that E is symmetric. Let Δ be a countable subgroup of X containing E and write $\Delta = \bigcup_{n=1}^{\infty} \Delta_n$ where $\{\Delta_n\}$ is an increasing sequence of finite symmetric sets and $1 \in \Delta_1$. Let m be as given by Lemma 8.9.

By that lemma we can write

$$E = F_1 \cup (\bigcup_{i \in I_1} F_{i,1}) \quad \text{(disjoint)} \tag{1_1}$$

where

$$F_1 \text{ and all } F_{i,1} \text{ are finite and symmetric,} \tag{2_1}$$

$$|F_{i,1}| \leqq m \text{ for all } i \in I_1, \tag{3_1}$$

and

$$F_{i,1}F_{j,1} \cap \Delta_1 = \emptyset \text{ for } i \neq j, \ i,j \in I_1. \tag{4_1}$$

By enlarging F_1, if necessary, we can also arrange for

$$F_1 \supset \Delta_1 \cap E. \tag{5_1}$$

Now we apply Lemma 8.9 to $E \setminus F_1$ to obtain

$$E \setminus F_1 = F_2 \cup (\bigcup_{i \in I_2} F_{i,2}) \quad \text{(disjoint)} \tag{1_2}$$

where

$$F_2 \text{ and all } F_{i,2} \text{ are finite and symmetric,} \tag{2_2}$$

$$|F_{i,2}| \leqq m \text{ for all } i \in I_2, \tag{3_2}$$

and

$$F_{i,2}F_{j,2} \cap \Delta_2 = \emptyset \text{ for } i \neq j, \ i,j \in I_2. \tag{4_2}$$

By enlarging F_2, if necessary, we can also arrange for

$$F_2 \cup F_1 \supset \Delta_2 \cap E \tag{5_2}$$

and

$$F_2 \cap F_{i,1} \neq \emptyset \text{ implies } F_{i,1} \subset F_2 \text{ for all } i \in I_1. \tag{6_2}$$

By induction and repeated use of Lemma 8.9, we obtain partitions

$$E \setminus (F_1 \cup F_2 \cup \cdots \cup F_{n-1}) = F_n \cup (\bigcup_{i \in I_n} F_{i,n}), \tag{1_n}$$

where

$$F_n \text{ and all } F_{i,n} \text{ are finite and symmetric,} \tag{2_n}$$

$$|F_{i,n}| \leqq m \quad \text{for all } i \in I_n, \tag{3_n}$$

$$F_{i,n}F_{j,n} \cap \Delta_n = \emptyset \quad \text{for } i \neq j, \quad i,j \in I_n, \tag{4_n}$$

$$F_n \cup F_{n-1} \cup \cdots \cup F_1 \supset \Delta_n \cap E, \tag{5_n}$$

and

$$F_n \cap F_{i,n-1} \neq \emptyset \text{ implies } F_{i,n-1} \subset F_n \quad \text{for } i \in I_{n-1}. \tag{6_n}$$

For each $n = 1,2,\ldots,$ let J_n denote the finite set $\{i \in I_n : F_{i,n} \subset F_{n+1}\}$. For each $i \in J_n$, write

$$F_{i,n} = \{\psi_{i,n,1}, \psi_{i,n,2}, \ldots, \psi_{i,n,m}\} \tag{7}$$

which is possible by (3_n) if we allow repetitions. Let

$$E_1 = F_1 \cup \bigcup_{n=1}^{\infty} \{\psi_{i,n,1} : i \in J_n\}$$

and for $2 \leqq k \leqq m$, let

$$E_k = \bigcup_{n=1}^{\infty} \{\psi_{i,n,k} : i \in J_n\}.$$

We need to show

$$E = E_1 \cup E_2 \cup \cdots \cup E_m \tag{8}$$

and

$$\text{each } E_k \text{ tends to infinity.} \tag{9}$$

To check (8), consider ψ in E. Since $E \subset \Delta$, (5_n) shows that $E = \bigcup_{n=1}^{\infty} F_n$. Let n_o be the least integer for which $\psi \in F_{n_o}$. If $n_o = 1$, then $\psi \in E_1$. If $n_o \geqq 2$, then (1_{n_o-1}) shows that $\psi \in F_{j,n_o-1}$ for some $j \in I_{n_o-1}$. Since ψ is in $F_{n_o} \cap F_{j,n_o-1}$, (6_{n_o}) implies that $F_{j,n_o-1} \subset F_{n_o}$

and so $j \in J_{n_o-1}$. Thus we have $\psi = \psi_{j,n_o-1,k}$ for some $k \in \{1,2,\ldots,m\}$ by (7) and hence $\psi \in E_k$. Hence (8) holds.

Consider k in $\{1,2,\ldots,m\}$. To check (9) it suffices to verify that E_k tends to infinity with respect to each Δ_{n_o}. In fact, we show that

$$\psi, \chi \in \bigcup_{n=n_o+1}^{\infty} \{\psi_{i,n,k} : i \in J_n\}, \quad \psi \neq \chi \quad \text{imply} \quad \psi\chi^{-1} \notin \Delta_{n_o}. \tag{10}$$

[Since each J_n is finite, we've only removed a finite set from E_k.] Assume that $\psi = \psi_{i,n,k}$, $\chi = \psi_{j,n',k}$, $\psi \neq \chi$ and $\psi\chi^{-1} \in \Delta_{n_o}$, where $i \in J_n$, $j \in J_{n'}$ and $n_o + 1 \leqq n \leqq n'$. Assume that $n < n'$. Then $\chi \in F_{j,n'} \subset F_{n'+1}$ and so by $(1_{n'+1})$,

$$\chi \in E \setminus (F_1 \cup F_2 \cup \cdots \cup F_{n'}). \tag{11}$$

Since $n < n'$, (1_n) shows that $\chi \in F_{i',n}$ for some $i' \in I_n$. We also have $\psi = \psi_{i,n,k} \in F_{i,n}$. If $i \neq i'$, then

$$\psi\chi^{-1} \in \Delta_{n_o} \cap F_{i,n}F_{i',n} \subset \Delta_n \cap F_{i,n}F_{i',n},$$

contrary to (4_n). Therefore $i = i' \in J_n$ and so

$$\chi \in F_{i',n} = F_{i,n} \subset F_{n+1},$$

contrary to (11). We conclude that $n < n'$ is impossible and so $n = n'$. Thus $\psi = \psi_{i,n,k}$ and $\chi = \psi_{j,n,k}$ where $i \neq j$. Hence

$$\psi\chi^{-1} \in \Delta_{n_o} \cap F_{i,n}F_{j,n} \subset \Delta_n \cap F_{i,n}F_{j,n},$$

which again contradicts (4_n). This proves (10). $\square$

9.2 REMARK. It is not known whether Theorem 9.1 holds for uncountable Sidon sets. If not, then some group X contains a Sidon set that is not a finite union of Sidon sets tending to infinity. Such a set cannot be a Stechkin set by 8.5(e), since any Stechkin set can be written as a finite union of asymmetric Rider sets. It is unknown whether every Sidon set is a Stechkin set.

The following assertion is immediate from 8.23 and 8.20: (a) If G is connected, if E is a Sidon set in X, and if f in C_E vanishes on a nonvoid open set, then $f \equiv 0$. Other results along these lines can be inferred from Déchamps-Gondim's theorems. However, as we observed in 8.24, her theorems are not valid if "every compact subset of G with nonvoid interior" is replaced by "every compact subset of G with positive Haar measure". To sidestep this difficulty we next introduce and study analogues of associated and strictly associated sets. We will then obtain improvements on the result (a) quoted above; see Corollaries 9.13 and 9.16.

9.3 DEFINITIONS. Let K be a measurable subset of G and E a subset of X. Then K and E are <u>strictly 2-associated</u> if there is a constant $\eta \geqq 1$ so that

$$\|f\|_2 \leqq \eta \|f \xi_K\|_2 \quad \text{for all} \quad f \in \text{Trig}_E(G). \tag{1}$$

If K and $E \setminus F$ are strictly 2-associated for some finite subset F of E, then K and E are said to be 2-<u>associated</u>. [Compare Definition 8.1.]

The results in 9.4 - 9.10 are due to Bonami [1970; Ch. IV] except for the nonconnected cases of 9.9 and 9.10; see López [1974]. Some of the results were announced by Bonami and Meyer [1969].

9.4 THEOREM. _Let_ E _be a_ $\Lambda(q)$ _set in_ X _for some_ $q > 2$, _and suppose that_ K _and_ E _are strictly_ 2-_associated for some measurable subset_ K _of_ G. _There exists_ ε _in_ $(0,1)$ _so that if_ K_o _is a measurable subset of_ K _and_ $\lambda(K \setminus K_o) < \varepsilon$, _then_ K_o _and_ E _are strictly_ 2-_associated_.

Proof. By Theorem 5.3 there is a constant η_q so that $\|f\|_q \leq \eta_q \|f\|_2$ for all $f \in \mathrm{Trig}_E(G)$. Let η be as in 9.3(1). Let $\alpha = 1 - \frac{2}{q}$ and choose $\varepsilon \in (0,1)$ so that

$$\varepsilon^{\alpha} \eta_q^2 \eta^2 < 1.$$

If $\lambda(K \setminus K_o) < \varepsilon$ and $f \in \mathrm{Trig}_E(G)$, then by Hölder's inequality, we have

$$\begin{aligned}\int_{K\setminus K_o} |f|^2 \, d\lambda &\leq \left(\int_{K\setminus K_o} 1 \, d\lambda\right)^{\alpha} \left(\int_{K\setminus K_o} |f|^{2\frac{q}{2}} \, d\lambda\right)^{2/q} \\ &\leq \lambda(K \setminus K_o)^{\alpha} \|f\|_q^2 \leq \varepsilon^{\alpha} \eta_q^2 \|f\|_2^2 \\ &\leq \varepsilon^{\alpha} \eta_q^2 \eta^2 \|f \xi_K\|_2^2 = \varepsilon^{\alpha} \eta_q^2 \eta^2 \int_K |f|^2 \, d\lambda .\end{aligned}$$

Hence we have

$$\int_{K_o} |f|^2 \, d\lambda \geq (1 - \varepsilon^{\alpha} \eta_q^2 \eta^2) \int_K |f|^2 \, d\lambda ,$$

that is,

$$(1 - \varepsilon^{\alpha} \eta_q^2 \eta^2)^{-\frac{1}{2}} \|f \xi_{K_o}\|_2 \geq \|f \xi_K\|_2 .$$

By 9.3(1) again we obtain

$$\eta(1 - \varepsilon^{\alpha} \eta_q^2 \eta^2)^{-\frac{1}{2}} \|f \xi_{K_o}\|_2 \geq \|f\|_2 .$$

Thus K_o and E are strictly 2-associated. □

It is clear from 9.3(1) that G and E are strictly 2-associated for all subsets E of X. Hence the next corollary is obvious.

9.5 COROLLARY. _If_ E _is a_ $\Lambda(q)$ _set in_ X _for some_ $q > 2$, _then there is_ $\varepsilon \in (0,1)$ _so that_ K _and_ E _are strictly_ 2-_associated for all measurable subsets_ K _of_ G _satisfying_ $\lambda(G \setminus K) < \varepsilon$.

9.6 THEOREM. _If_ E _is a_ $\Lambda(4)$ _set in_ X _that tends to infinity, then_ K _and_ E _are_ 2-_associated for all measurable subsets_ K _of_ G _of positive measure_.

Proof. Choose $\kappa \geqq 1$ so that $\|f\|_4 \leqq \kappa\|f\|_2$ for all f in $\text{Trig}_E(G)$. Let K be a measurable set in G with $\lambda(K) > 0$. Since the characteristic function ξ_K belongs to $L^2(G)$, its Fourier transform is in $\ell^2(X)$; so there is a finite set $\Delta \subset X$ such that

$$\Big(\sum_{\chi\in X\setminus\Delta} |\hat{\xi}_K(\chi^{-1})|^2\Big)^{\frac{1}{2}} < \frac{\lambda(K)}{2\kappa^2} . \tag{1}$$

We may assume that Δ contains the character 1. Since E tends to infinity, E contains a finite subset F satisfying

$$\chi, \psi \in E \setminus F \text{ and } \chi \neq \psi \text{ imply } \chi\psi^{-1} \notin \Delta. \tag{2}$$

Let $f \in \text{Trig}_{E\setminus F}(G)$ and consider $g = |f|^2 = f\cdot\bar{f} \in \text{Trig}(G)$. For χ in X we have

$$\hat{g}(\chi) = \hat{f}*\hat{\bar{f}}(\chi) = \sum_{\psi\in X} \hat{f}(\psi)\hat{\bar{f}}(\psi^{-1}\chi) = \sum_{\psi\in X} \hat{f}(\psi)\overline{\hat{f}(\psi\chi^{-1})}.$$

If χ is in $\text{Supp}(\hat{g})$, then for some ψ in X, both ψ and $\psi\chi^{-1}$ belong to $\text{Supp}(\hat{f}) \subset E \setminus F$. If also $\chi \neq 1$, then (2) shows that $\chi = \psi(\psi\chi^{-1})^{-1} \notin \Delta$. That is,

$$\text{Supp}(\hat{g}) \subset (X \setminus \Delta) \cup \{1\}.$$

Now observe that

$$\int_K |f|^2 d\lambda = \int_K \sum_{\chi\in X} \hat{g}(\chi)\chi\, d\lambda = \sum_{\chi\in X} \hat{g}(\chi)\xi_K^{\wedge}(\chi^{-1})$$
$$= \hat{g}(1)\xi_K^{\wedge}(1) + \sum_{\chi\in X\setminus\Delta} \hat{g}(\chi)\xi_K^{\wedge}(\chi^{-1})$$
$$= \|f\|_2^2\,\lambda(K) + \sum_{\chi\in X\setminus\Delta} \hat{g}(\chi)\xi_K^{\wedge}(\chi^{-1}). \qquad (3)$$

Hölder's inequality and (1) show that the last sum is bounded in modulus by

$$\|\hat{g}\|_2\Big(\sum_{\chi\in X\setminus\Delta} |\xi_K^{\wedge}(\chi^{-1})|^2\Big)^{\frac{1}{2}} \leqq \|g\|_2 \frac{\lambda(K)}{2\kappa^2} = \|f\|_4^2 \frac{\lambda(K)}{2\kappa^2}$$
$$\leqq \kappa^2\|f\|_2^2 \frac{\lambda(K)}{2\kappa^2} = \tfrac{1}{2}\|f\|_2^2\,\lambda(K) .$$

From this information and (3) we infer that

$$\int_K |f|^2 d\lambda \geqq \tfrac{1}{2}\lambda(K)\,\|f\|_2^2.$$

Thus $E\setminus F$ and K are strictly 2-associated [indeed with $\eta = \sqrt{2}\,\lambda(K)^{-\frac{1}{2}}$] and so E and K are 2-associated. □

Note that Theorem 9.6 applies to all dissociate sets and many Rider sets; see 8.5(d) and 8.5(e). The situation for more general Sidon sets is clarified by Theorem 9.11.

The next lemma is similar to, but easier than, Lemma 8.20. We omit its proof, which uses nothing more sophisticated than the open mapping theorem.

9.7 LEMMA. _Let_ E _be a subset of_ X _and_ K _a measurable subset of_ G _such that_ $\lambda(K) > 0$. _Then_ E _and_ K _are strictly 2-associated if and only if the following two conditions hold:_

(i) $L_E^2(K) \equiv \{f|_K : f\in L_E^2(G)\}$ _is closed in_ $L^2(K)$;

(ii) $\rho(f) = f|_K$ _defines a one-to-one map on_ $L_E^2(G)$.

9.8 THEOREM. *Let* E *be a* $\Lambda(q)$ *set in* X *for some* $q > 2$ *and* K *a measurable set in* G *such that* $\lambda(K) > 0$. *Then* E *and* K *are strictly* 2-*associated if and only if*

(i) E *and* K *are* 2-*associated,*

and

(ii) $f \in \mathrm{Trig}_E(G)$ *and* $f|_K = 0$ λ-*a.e. imply* $f \equiv 0$.

Proof. If E and K are strictly 2-associated, then (i) and (ii) clearly hold.

Suppose that (i) and (ii) hold. Then $E = E_0 \cup \{\chi_1, \ldots, \chi_n\}$ where E_0 and K are strictly 2-associated. Since the hypotheses (i) and (ii) hold for each $E_k = E_0 \cup \{\chi_1, \ldots, \chi_k\}$, $1 \leq k \leq n$, a simple induction argument shows that it suffices to prove that E_1 and K are strictly 2-associated. In other words, we may assume that E itself has the form $E_0 \cup \{\psi\}$ where $\psi \notin E_0$ and E_0 and K are strictly 2-associated.

By Lemma 9.7, $L^2_{E_0}(K)$ is a closed subspace of $L^2(K)$. Since $L^2_E(K)$ is the sum of $L^2_{E_0}(K)$ and the one-dimensional subspace generated by $\psi|_K$, $L^2_E(K)$ is also closed in $L^2(K)$. By 9.7, then, it suffices to show that the restriction map ρ on $L^2_E(G)$ is one-to-one. Consider f in $L^2_E(G)$ where $f|_K = 0$ λ-a.e. Since E_0 and K are strictly 2-associated, Theorem 9.4 provides an ε in $(0,1)$ such that E_0 and K_0 are strictly 2-associated whenever $K_0 \subset K$ and $\lambda(K \setminus K_0) < \varepsilon$. Now let $K' = \{x \in K : f(x) = 0\}$ and for w in G let $K_w = (K' - w) \cap K'$. Then $\lambda(K_w) = \xi_K * \xi_{-K}(w)$ is a continuous function of w on G, and so there is a neighborhood W of

0 such that $\lambda(K \setminus K_w) < \varepsilon$ for all $w \in W$. Hence

$$E_0 \text{ and } K_w \text{ are strictly 2-associated for } w \in W. \tag{1}$$

For each w in W, define $g_w(x) = f(x+w) - \psi(w)f(x)$ for $x \in G$. Then $\hat{g}_w(\chi) = [\chi(w) - \psi(w)]\hat{f}(\chi)$ for all $\chi \in X$, and so each g_w belongs to $L^2_{E_0}(G)$. Note that g_w vanishes on K_w and so $g_w \equiv 0$ for $w \in W$ by (1). In particular, $\hat{g}_w \equiv 0$ for $w \in W$. Now if $\hat{f}(\chi) \neq 0$, then $\chi(w) = \psi(w)$ for $w \in W$. It follows that χ and ψ agree on the smallest (open) subgroup G_0 of G containing W, and so $\chi \in \psi X_0$ where X_0 is the annihilator of G_0. Since X_0 is finite, we conclude that $\mathrm{Supp}(\hat{f})$ is also finite. Hence f belongs to $\mathrm{Trig}_E(G)$ and hypothesis (ii) shows that $f \equiv 0$. This shows that c is one-to-one and completes the proof. □

9.9 COROLLARY. *Let E be a $\Lambda(q)$ set in X for some $q > 2$, and let K be a measurable subset of G so that $\lambda(K) > 0$ and E and K are 2-associated. If E is X_0-subtransversal for all finite subgroups X_0 of X, then E and K are strictly 2-associated. In particular, E and K are strictly 2-associated if G is connected.*

Proof. Property 9.8(ii) holds by Corollary 8.12. Note that if G is connected, then E is automatically X_0-subtransversal for all finite subgroups X_0 of X. □

9.10 COROLLARY. *Let E be a $\Lambda(4)$ set in X that tends to infinity and is also X_0-subtransversal for all finite subgroups X_0 of X. If $f \in L^2_E(G)$ vanishes on a set of positive measure, then $f = 0$ λ-a.e. In particular, if G is connected and $f \in L^2_E(G)$ vanishes on a set of positive measure, then $f = 0$ λ-a.e.*

Proof. Let $K = \{x \in G : f(x) = 0\}$. By Theorem 9.6, K and E are 2-associated. By Corollary 9.9, K and E are in fact strictly 2-associated. Hence $f = 0$ λ-a.e.; see, for example, 9.7(ii). □

We next investigate the relationship between the concepts just studied and Sidon sets.

9.11 THEOREM. For a Sidon set E in X, the following are equivalent:

(i) E is 2-associated with all measurable subsets K of G having positive measure.

(ii) E is almost X_0-subtransversal for all finite subgroups X_0 of X.

Proof. Suppose (i) holds and consider a finite subgroup X_0 of X. Its annihilator G_0 is an open subgroup of G. Hence by (i) there is a finite subset F of E and $\eta \geqq 1$ so that

$$\|f\|_2 \leqq \eta \|f \xi_{G_0}\|_2 \quad \text{for all} \quad f \in \mathrm{Trig}_{E\setminus F}(G). \tag{1}$$

In particular, if χ_1 and χ_2 are distinct characters in $E \setminus F$, and if we let $f = \chi_1 - \chi_2$ in (1), we see that the function $\chi_1 - \chi_2$ is not identically zero on G_0. Therefore $\chi_1\chi_2^{-1} \notin X_0$. This shows that $E \setminus F$ is X_0-subtransversal and so E is almost X_0-subtransversal.

Now suppose that (ii) holds and consider a measurable set K in G with positive measure. Let m be the positive integer given by Corollary 8.10 for the Sidon set E. Applying Lemma 8.14 to this m and $g = \xi_K$, we obtain a finite subgroup X_0 of X and $\varepsilon > 0$ such that

$$\|f\,\xi_K\|_2^2 \geqq 3\varepsilon\|f\|_2^2 \tag{2}$$

whenever $f \in \mathrm{Trig}(G)$, $\mathrm{length}(f) \leqq m$ and $\mathrm{Supp}(\hat{f})$ is X_0-subtransversal. Since E is a $\Lambda(4)$ set by Theorem 5.8, there exists $\varkappa > 0$ such that

$$\|f\|_4 \leqq \varkappa\|f\|_2 \quad \text{for } f \in \mathrm{Trig}_E(G). \tag{3}$$

There is a finite set $\Delta \subset X$ so that

$$\sum_{\chi\in X\setminus\Delta} |\xi_K^{\wedge}(\chi^{-1})|^2 < \varepsilon^2\varkappa^{-4}. \tag{4}$$

By Corollary 8.10, E can be written as a disjoint union $F \cup (\bigcup_{i\in I} F_i)$ where F is finite, $|F_i| \leqq m$ for all $i \in I$, and $F_iF_j^{-1} \cap \Delta = \emptyset$ for $i \neq j$, $i,j \in I$. By taking F a little larger, if necessary, we may suppose that $E\setminus F$ is X_0-subtransversal for the fixed X_0 obtained from 8.14; see (ii).

We prove that $E\setminus F$ and K are strictly 2-associated. Given f in $\mathrm{Trig}_{E\setminus F}(G)$, we write $f = \sum_{i\in I} f_i$ where each f_i belongs to $\mathrm{Trig}_{F_i}(G)$. Let $g = \sum_{i\neq j} f_i\overline{f_j}$ and observe that $|f|^2 = \sum_{i\in I}|f_i|^2 + g$. Using (3) we obtain

$$\begin{aligned}\|g\|_2 &= \Big\|\,|f|^2 - \sum_{i\in I}|f_i|^2\Big\|_2 \leqq \|f^2\|_2 + \sum_{i\in I}\|f_i^2\|_2 \\ &= \|f\|_4^2 + \sum_{i\in I}\|f_i\|_4^2 \leqq \varkappa^2\Big[\|f\|_2^2 + \sum_{i\in I}\|f_i\|_2^2\Big] \\ &= 2\varkappa^2\|f\|_2^2.\end{aligned} \tag{5}$$

A simple argument (compare the proof of 9.6) shows that if $\chi \in \mathrm{Supp}(\hat{g})$, then $\chi \in F_iF_j^{-1}$ for some $i \neq j$ and so $\chi \notin \Delta$. Thus $\mathrm{Supp}(\hat{g}) \subset X\setminus\Delta$ and (5) and (4) lead to

$$\begin{aligned}\Big|\int_K g\,d\lambda\Big| &= \Big|\sum_{\chi\in X\setminus\Delta}\hat{g}(\chi)\xi_K^{\wedge}(\chi^{-1})\Big| \leqq \|g\|_2\Big(\sum_{\chi\in X\setminus\Delta}|\xi_K^{\wedge}(\chi^{-1})|^2\Big)^{\frac12} \\ &\leqq 2\varkappa^2\|f\|_2^2\,\varepsilon\,\varkappa^{-2} = 2\varepsilon\|f\|_2^2.\end{aligned} \tag{6}$$

Since (2) applies to each f_i, we have

$$3\varepsilon\|f\|_2^2 = 3\varepsilon \sum_{i\in I} \|f_i\|_2^2 \leq \sum_{i\in I} \|f_i \xi_K\|_2^2$$

and so, using (6), we see that

$$\|f \xi_K\|_2^2 = \sum_{i\in I} \|f_i \xi_K\|_2^2 + \int_K g\, d\lambda \geq 3\varepsilon\|f\|_2^2 - 2\varepsilon\|f\|_2^2 = \varepsilon\|f\|_2^2.$$

This inequality holds for all f in $\mathrm{Trig}_{E\setminus F}(G)$ and so $E \setminus F$ and K are strictly 2-associated. □

9.12 COROLLARY. <u>For a Sidon set</u> E <u>in</u> X, <u>the following are equivalent</u>:

(i) E <u>is strictly</u> 2-<u>associated with all measurable subsets</u> K <u>of</u> G <u>of positive measure</u>.

(ii) E <u>is</u> X_0-<u>subtransversal for all finite subgroups</u> X_0 <u>of</u> X.

<u>Proof</u>. The proof of (i) ⇒ (ii) is the same as in 9.11; simply take $F = \emptyset$. If (ii) holds, then 9.11(i) holds and so (i) holds by Corollary 9.9. □

The next corollary is due to Bonami; see Déchamps-Gondim [1972; Remarque 6.2]. Bonami kindly supplied us with a detailed proof; 9.11 and 9.12 are based on it. See López [1974].

9.13 COROLLARY [BONAMI]. <u>If</u> G <u>is connected and</u> E <u>is a Sidon set in</u> X, <u>then</u> E <u>is strictly</u> 2-<u>associated with all measurable subsets</u> K <u>of</u> G <u>of positive measure</u>. <u>In particular, if</u> f <u>in</u> $L_E^2(G)$ <u>vanishes on a set of positive measure, then</u> $f = 0$ λ-<u>a.e.</u>

The converse of Corollary 9.13 can fail dramatically. That is, many non-Sidon sets are strictly 2-associated with all measurable sets of positive measure. The last three theorems in Bonami [1970] illustrate this. For example, she shows that if E is a Hadamard set in $\mathbb{Z}$ with lacunarity constant ≥ 3 [see 2.10(1)], and if E_k is defined as in 5.13, then E_k is strictly 2-associated with all sets in $\mathbb{T}$ of positive measure. As noted in Corollary 5.14, such sets cannot be Sidon sets unless, of course, $k = 1$.

We next give some routine, but interesting, consequences of the preceding results. For this purpose, we again consider an increasing sequence $\{X_n\}_{n=1}^{\infty}$ of finite subsets of X, and we write X_∞ for $\bigcup_{n=1}^{\infty} X_n$. The symbols $s_n f$ and $s_n \phi$ are also defined as in 7.12. The proof of the next lemma is similar to that of Theorem 7.13; we omit it.

*9.14 LEMMA. <u>The following are equivalent for subsets</u> E <u>of</u> X_∞ <u>and measurable subsets</u> K <u>of</u> G.

(i) E <u>and</u> K <u>are</u> 2-<u>associated</u>.

(ii) <u>For each complex-valued function</u> ϕ <u>on</u> E,

$$\sup_n \|(s_n\phi)\xi_K\|_2 < \infty \quad \text{implies} \quad \phi \in \ell^2(E).$$

*9.15 THEOREM. <u>Let</u> E <u>be a subset of</u> X_∞. <u>The following three properties are equivalent</u>:

(i) E <u>is a Sidon set and is almost</u> X_0-<u>subtransversal for all finite subgroups</u> X_0 <u>of</u> X.

(ii) <u>If</u> ϕ <u>is a complex-valued function on</u> E <u>and</u> $\sup_n |s_n\phi|$ <u>is bounded on some nonvoid open subset of</u> G, <u>then</u> ϕ <u>belongs to</u> $\ell^1(E)$.

(iii) <u>If</u> ϕ <u>is a complex-valued function on</u> E <u>and</u> $\sup_n |s_n\phi(x)| < \infty$ <u>for each</u> x <u>in some nonvoid open set in</u> G, <u>then</u> ϕ <u>belongs to</u> $\ell^1(E)$.

These properties imply

(iv) *If* ϕ *is a complex-valued function on* E *and* $s_n\phi(x)$ *converges on a nonvoid open set, then* $\phi \in \ell^1(E)$.

(v) *If* ϕ *is a complex-valued function on* E *and* $\sup_n |s_n\phi(x)| < \infty$ *for all* x *in a set of positive measure, then* $\phi \in \ell^2(E)$.

(vi) *If* ϕ *is a complex-valued function on* E *and* $s_n\phi(x)$ *converges on a set of positive measure, then* $\phi \in \ell^2(E)$.

Proof. *(i) ⇒ (ii)*. If $\sup_n |s_n\phi|$ is bounded on a nonvoid open set, then $\sup_n \|(s_n\phi)\xi_K\|_u < \infty$ for some compact set K with nonvoid interior. By Corollary 8.18, E is associated with K and so $E \setminus F$ is strictly associated with K for some finite subset F of E. If $\phi_1 = \phi\xi_{E\setminus F}$, then we have $\sup_n \|(s_n\phi_1)\xi_K\|_u < \infty$. It follows from 8.1(1) that $\sup_n \|s_n\phi_1\|_A < \infty$. Hence ϕ_1 and ϕ belong to $\ell^1(E)$.

(ii) ⇒ (iii). Suppose that $\sup_n |s_n\phi(x)| < \infty$ for each x in the nonvoid open set W. For each $m \in \mathbb{Z}^+$, $W_m = \{x \in W : \sup_n |s_n\phi(x)| \leq m\}$ is a relatively closed set in W. By the Baire category theorem, some W_m has interior in W and hence in G. So ϕ is in $\ell^1(E)$ by (ii).

(iii) ⇒ (ii). Obvious.

(ii) ⇒ (i). Let K be any compact subset of G with nonvoid interior. If (ii) holds, then 7.15(iii) holds. Hence 7.15(i) holds, i.e. E and K are associated. Since K is arbitrary, (i) holds by Corollary 8.18.

(iii) ⇒ (iv). Obvious.

<u>(i) ⇒ (v)</u>. Let $K = \{x \in G : \sup_n |s_n\phi(x)| < \infty\}$, and for each $m \in \mathbb{Z}^+$, let $K_m = \{x \in G : \sup_n |s_n\phi(x)| \leq m\}$. Since $\lambda(K) > 0$, we have $\lambda(K_m) > 0$ for some m. It is obvious that $\sup_n \|(s_n\phi)\xi_{K_m}\|_2 < \infty$ and so E and K_m are 2-associated by Theorem 9.11. Hence $\phi \in \ell^2$ by Lemma 9.14.

<u>(v) ⇒ (vi)</u>. Obvious. □

*9.16 COROLLARY. <u>Suppose that $E \subset X_\infty$ and that G is connected. Then E is a Sidon set if and only if 9.15(ii) holds, and also if and only if 9.15(iii) holds. If E is a Sidon set, then 9.15(iv) - (vi) hold, and also</u>

(i) <u>if ϕ is a complex-valued function on E and $s_n\phi(x) \to 0$ on a set of positive measure, then $\phi \equiv 0$.</u>

<u>Proof</u>. Only (i) requires some comment. If $s_n\phi(x) \to 0$ on a set of positive measure, then $\phi \in \ell^2(E)$ by 9.15(vi). Thus we have $\phi = \hat{f}|_E$ for some $f \in L^2_E(G)$. Since we have $\lim_n \|s_n\phi - f\|_2 = 0$, a subsequence of $\{s_n\phi\}$ converges λ-a.e. to f. It follows that f vanishes on a set of positive measure. Now Corollary 9.13 tells us that $f = 0$ λ-a.e. and hence $\phi \equiv 0$. □

The following theorem is similar to the equivalences of (i) - (iii) in 9.15.

*9.17 THEOREM. <u>Suppose that the sets X_n and X_∞ are symmetric and consider a symmetric subset E of X_∞. The following are equivalent:</u>

(i) <u>E is a full FZ-set.</u>

(ii) <u>If ϕ is hermitian on E and $\sup_n s_n\phi$ is bounded above on some nonvoid open set, then $\phi \in \ell^1(E)$.</u>

(iii) _If_ ϕ _is hermitian on_ E _and_ $\sup_n s_n\phi(x) < \infty$ _for each_ x _in some nonvoid open set, then_ $\phi \in \ell^1(E)$.

Proof. Use Theorem 7.13 to prove the equivalence of (i) and (ii), and use the Baire category theorem to prove the implication (ii) ⇒ (iii). □

***9.18 REMARK.** **Zygmund [1930], [1932] proved 9.15(vi) for Hadamard sets; see [Z; Vol. I, page 203, 6.4]. Related results for dissociate sets are given by Hewitt and Zuckerman [1966; 4.1,4.2]. Zygmund [1948] established 9.16(i) for Hadamard sets; see [Z; Vol. I, page 206, 6.13(ii)]. Zygmund [1931] established 9.17(iii) for Hadamard sets ([Z; Vol. I, page 247, 6.3]) and Gaposhkin [1967b] generalized the result to Stechkin sets in $\mathbb{Z}$. He also proved an analogue for Walsh systems which is equivalent to the following: For symmetric Stechkin sets in the dual of $\mathbb{Z}(2)^{\aleph_0}$, 9.17(iii) holds if and only if the set tends to infinity. A discussion of these matters also appears in [HR; Vol. II, pp. 448-449].**

A very special case of Corollary 9.13 is established by Orlov [1973].

Chapter 10

SELECTED GUIDE TO LACUNARITY

For the early history of lacunarity, see the Notes to Section 37 in [HR; Vol. II] and Kahane [1964]. The following books also deal with lacunarity: [DR; Chapter 5], [E; Chapter 15], [K; Chapter 10], [Ka; Chapter VI], [Kz; Chapter 5], [LP; Chapters 5 and 6], [R; Section 5.7], and [Z; Vol. I, pp. 202-212, 247-249 and Vol. II, pp. 131-133].

10.1 OTHER CHARACTERIZATIONS OF SIDON SETS. Ramirez [1968] characterized Sidon sets in terms of the natural pairing between $M(G)^\wedge$ and $M(X)$; see [DR; page 49]. Further investigations in this direction were made by Hartman [1972]. The implication 1.3(iii) ⇒ 1.3(i) is shown by Badé and Curtis [1966; Corollary 3.4] to be a special case of a much more general phenomenon.

A lemma in Graham [1973b] shows that a set E in X is a Sidon set if and only if there exists a discontinuous complex-valued function on $[-1,1]$ that operates in the algebra $\{\hat{\mu}|_E : \mu \in M(G)\}$.

Dunkl and Ramirez [1973] show that if E is an infinite Sidon set in X, then there does not exist a bounded linear map $\pi : c_o(E) \to L^1(G)$ satisfying $(\pi(\phi))^\wedge|_E = \phi$ for all ϕ in $c_o(E)$. Similarly, there does not exist a bounded linear map $\pi : \ell^\infty(E) \to M(G)$ satisfying $(\pi(\phi))^\wedge|_E = \phi$ for all ϕ in $\ell^\infty(E)$.

10.2 RIESZ PRODUCTS. Theorem 2.7 is used by Hewitt and Zuckerman [1966] to construct singular measures on compact abelian groups with absolutely continuous convolution squares. Stromberg [1968] constructs large independent sets of such measures. Some recent interesting uses of Riesz products appear in Padé [1973], Peyriere [1973], Graham [1974] and Brown [1974]. Two useful survey articles on Riesz products are Hewitt [1968] and Keogh [1965].

Riesz products for noncommutative groups are developed by Parker [1972] and Cygan [1974].

10.3 $\Lambda(p)$ SETS. The texts listed at the beginning of this chapter deal with $\Lambda(p)$ sets as well as with Sidon sets. The terminology originated with Rudin [1960], who established many of the basic properties of Sidon and $\Lambda(p)$ sets in $\mathbb{Z}$.

Consider $1 < p < \infty$. The tensor product of $L^p(G)$ and $L^{p'}(G)$ can be realized as a space $A_p(G)$ of continuous functions on G. The conjugate space of $A_p(G)$ is isometrically isomorphic with the space $M_p(G)$ of all multipliers of $L^p(G)$ into $L^p(G)$ that commute with translation; $M_p(G)$ can be regarded as a subspace of $\ell^\infty(X)$. These facts are established by Figà-Talamanca [1964], [1965]. He also observes [1964; Theorem 7] that the following are equivalent for a subset E of X and $p > 2$:

(i) E is a $\Lambda(p)$ set;

(ii) $(A_p(G))_E \subset A(G)$;

(iii) each ϕ in $\ell^\infty(E)$ is the restriction to E of an element in $M_p(G)$;

(iv) each ϕ in $\ell^\infty(E)$ is the restriction to E of

an element in $M_p(G)$ that vanishes outside of E.
See also Figà-Talamanca and Rider [1966; Theorem 6]. Figà-Talamanca [1964] calls sets satisfying (ii) "p-Sidon sets", as does Stafney [1969], but we use this term for a somewhat different notion; see 10.6.

Bachelis and Ebenstein [1974] have shown that if E is a $\Lambda(p)$ set where $1 \leq p < 2$, then E is a $\Lambda(p+\varepsilon)$ set for some $\varepsilon > 0$. Bachelis [1971] characterized $\Lambda(p)$ sets for p an even integer, and Ebenstein [1974] obtained the same characterization for all $p > 2$. The problem of finding $\Lambda(p)$ sets that are not $\Lambda(q)$ sets, $p < q$, has been dealt with by Bonami [1970; p. 355], Edwards, Hewitt and Ross [1972a], and Benke [1974b].

Notions that appear very similar to $\Lambda(p)$ sets are investigated by Gaposhkin [1967a], Fournier [1974; section 3], Fournier [1973a], and Dressler, Parker and Pigno [1973]. The last paper deals with "small p sets" and Sidon sets.

Edwards and Ross [1973] calculate and estimate various $\Lambda(2)$ constants and Sidon constants. They observe that if E is a Sidon set with Sidon constant $\varkappa$, then the $\Lambda(2)$ constant for E cannot exceed $\sqrt{2}\,\varkappa$. In particular, the $\Lambda(2)$ constant for the set of projections of $G = \mathbb{T}^{\aleph_0}$ does not exceed $\sqrt{2}$.

The following is still an open question. Let E be a $\Lambda(4)$ set in the character group of $\mathbb{Z}(p)^{\aleph_0}$. Must E be a $\Lambda(q)$ set for some $q > 4$? Benke [1974b] shows that such sets need not be $\Lambda(q)$ sets for any $q > 4p$.

10.4 ROSENTHAL SETS. Theorem 1.3 shows that Sidon sets E satisfy the equality $L_E^{\infty}(G) = C_E(G)$. Rosenthal [1967] (see [HR; 37.25]) proved that there are sets $E \subset \mathbb{Z}$ such that

$L_E^\infty(G) = C_E(G)$ that are not Sidon sets. In fact, he constructs such sets containing arbitrarily long arithmetic progressions (compare 6.8). Rosenthal's proof relied heavily on properties of $\mathbb{R}$ and $\mathbb{T}$. Blei [1972b] ingeniously extended Rosenthal's result to all compact infinite abelian groups: there always exist non-Sidon sets E in X satisfying $L_E^\infty(G) = C_E(G)$. Indeed, every non-Sidon set contains such a set. Refinements of these results appear in Blei [1974a]. See also Blei [1974b].

Pigno and Saeki [1974] have extended Rosenthal's original examples as follows. For $n = 1,2,\ldots,$ let $E_n = \{0,\pm 1,\ldots,\pm n\}$ and let $(a_0,a_1,a_2,\ldots)$ be a sequence of integers $\geqq 2$. Then the set

$$E = \bigcup_{n=1}^{\infty} a_0 a_1 \cdots a_{n-1} E_n$$

is a Rosenthal set and so is any finite union of translates of such sets.

10.5 RIESZ SETS. From 5.3(ii) it follows that Sidon sets and $\Lambda(p)$ sets satisfy the relation $M_E(G) = L_E^1(G)$. The set $\mathbb{Z}^+$ in $\mathbb{Z}$ also satisfies this relation (with $G = \mathbb{T}$) by the F. and M. Riesz theorem [R; 8.2.1]. For this reason, sets satisfying $M_E(G) = L_E^1(G)$ are sometimes called Riesz sets. Meyer [1968b] studied such sets in detail. He also called a set E in X a strong Riesz set if its closure in the topology inherited from the Bohr compactification bX is a Riesz set. Meyer proved that the union of a Riesz set and a strong Riesz set is again a Riesz set. He also proved that the set of perfect squares in $\mathbb{Z}^+$ and the set of primes in $\mathbb{Z}^+$ are both strong Riesz sets. These results were extended by Dressler and Pigno [1974a], [1974c].

Pigno and Saeki [1973] have shown that the union of a Sidon set and a Riesz set is a Riesz set. Dressler and Pigno [1974b] establish the following extension of the F. and M. Riesz theorem: If $E \subset \mathbb{Z}$ is a Rosenthal set, that is $L_E^{\infty}(\mathbb{T}) = C_E(\mathbb{T})$, then $E \cup \mathbb{Z}^+$ is a Riesz set. In [1975] they show that the union of a Riesz set and a Rosenthal set is a Riesz set. Extensions of the F. and M. Riesz theorem involving $\Lambda(1)$ sets appear in Rudin [1960; 5.7] and in Pigno [1974b]. Riesz sets in general LCA groups are investigated by Pigno [1974a].

10.6 p-SIDON SETS. According to Theorem 1.3, Sidon sets E are precisely those sets satisfying

$$f \in C_E(G) \quad \text{implies} \quad \hat{f} \in \ell^1(X).$$

Bożejko and Pytlik [1972], Hahn [1972], and Edwards and Ross [1974] studied sets E satisfying

$$f \in C_E(G) \quad \text{implies} \quad \hat{f} \in \ell^p(X), \tag{1}$$

where $1 \leqq p < 2$. The latter authors termed such sets p-Sidon sets. They showed that there are 4/3-Sidon sets that are not Sidon sets; compare Blei [1974b]. This last result has been generalized by Johnson and Woodward [1974]. Very recent work on p-Sidon sets appears in Johnson [1974] and Woodward [1974]. For each p, $1 < p < 2$, Suslikova [1973] constructs a set $E \subset \mathbb{Z}$ that is a UC-set, but not a p-Sidon set. More is possible, since Theorem 3 of Pedemonte [1974] and Corollary 8(2) of Johnson [1974] together show that $\mathbb{Z}$ contains a UC-set that cannot be written as a finite union of p-Sidon sets.

Compact analogues to p-Sidon sets are the p-Helson sets studied by Gregory [1972].

10.7 MISCELLANEA. Lacunary sets are often used to construct multipliers with certain properties. For examples, see [HR; 37.22,37.24], Bonami [1970], Fournier [1973b; Theorem 3], Meyer [1968a; Chap. 4], Price [1970; section 3] and Wells [1971].

Rider [1966b] defines Bohr sets in $\mathbb{Z}^+$ in terms of Dirichlet series. He characterizes the Bohr sets in $\mathbb{Z}^+$ in terms of certain Sidon sets in $\mathbb{Z}^{\aleph_0}$. Fournier [1969; section 5] modified Rider's definition and obtained necessary arithmetic conditions on Bohr sets. He also studied Sidon sets and $\Lambda(2)$ sets in connection with convex sets in X.

Edwards [1973] studies functions on $\mathbb{T}$ that belong to various Lipschitz classes and whose Fourier series are E-spectral for some Sidon set E. See also Benke [1974d].

Sidon sequences in normed vector spaces and in quotient algebras of group algebras are studied by Meyer [1972].

10.8 UNION THEOREMS. Varopoulos modified the techniques of Drury in Chapter 3 to settle in the affirmative a big unsolved problem: Is the finite union of Helson sets another Helson set? The "right" proof of this result is given by Herz [1972]. For the history of the problem and other references, see McGehee's excellent review of Herz's paper, Math. Reviews 46 (1973), review 5939.

10.9 FOURIER-STIELTJES TRANSFORMS OF CONTINUOUS MEASURES. Hartman and Ryll-Nardzewski [1971] asked whether a continuous measure μ in $M(G)$ can satisfy $\inf\{|\hat{\mu}(\chi)| : \chi \in E\} > 0$ for a non-Sidon set E; compare Corollary 4.2. Kaufman [1971] showed that the answer is yes. Graham [1973a] and Ramirez [1973] have given constructive proofs. See also Méla [1968a; section 6].

Chaney [1971] constructed a set E in $\mathbb{Z}$ such that all E-spectral measures on $\mathbb{T}$ are continuous, but not all of them are absolutely continuous. Saeki has posed the following problem: Suppose $\mu \in M(G)$ and $\hat{\mu} \in c_0(X)$ imply that $\hat{\mu}|_E = \hat{f}|_E$ for some $f \in L^1(G)$. Must E be a Sidon set? See Pigno and Saeki [1973]. Continuous measures and analogues to Wiener's Lemma 4.1 are also studied by Baker [1972].

10.10 INTERPOLATION SETS. A set E in X is called a set of interpolation if every ϕ in $\ell^\infty(E)$ has the form $\hat{\mu}|_E$ for a discrete measure μ on G. Kahane [1966] proved that this is equivalent to: Every ϕ in $\ell^\infty(E)$ can be extended to an almost periodic function on X. Even if X is merely an LCA group, such sets are called I_0-sets. That is, $E \subset X$ is an I_0-set if <u>every</u> bounded function on E can be extended to a continuous almost periodic function on X. The set E is called an I-set if every bounded function on E that is uniformly continuous on E can be extended to a continuous almost periodic function on X.

There is a fairly vast literature on interpolation sets, some of which is very closely related to the theory of Sidon sets. Clearly sets of interpolation are Sidon sets; H. Rosenthal and W. Rudin have shown that the converse is in general false (Kahane [1966]). Strzelecki [1963] proved that Hadamard sets in $\mathbb{Z}^+$ are sets of interpolation; see also Méla [1964] and [1969; pp. 36-39], and [K; pp. 129-131]. Other papers on interpolation sets are Hartman [1961], [1968], [1972a], [1974], Ryll-Nardzewski [1963] and [1964], Hartman and Ryll-Nardzewski [1963], [1964], [1966] and [1967], Hartman, Kahane and Ryll-

Nardzewski [1965], Hartman, Rolewicz and Ryll-Nardzewski [1967], and Méla [1968b]. Important papers dealing with both interpolation sets and lacunarity are Méla [1968a] and [1969]. See also Blei [1974c].

10.11 ASSOCIATED SETS. Let G be an LCA group with character group X. Déchamps-Gondim's theorems in Chapter 8 and Bonami's theorems in Chapter 9 are instances where sets in X or spaces of functions on G are studied via a norm or semi-norm given by $\|f \xi_K\|$; here K is a subset of G and $\| \ \|$ is some norm.

For example, let $E = \{\lambda_j\}_{j=1}^{\infty}$ be an increasing sequence in $\mathbb{R}$ and let K be a compact subset of $\mathbb{R}$. Helson and Kahane [1965] say that K is appropriate to E if, for some $k \in \mathbb{Z}^+$, the norms $\|f \xi_K\|_u$ and $\sum_{j=k}^{\infty} |a_j|$ are equivalent for the set of trigonometric polynomials $f(x) = \sum_{j=k}^{\infty} a_j \exp(\lambda_j x)$; see also Kaufman [1969]. This concept is studied in the setting of compact abelian groups by Méla [1964] and in the setting of LCA groups by Méla [1968a] and Baker [1970]. Similar notions that are also termed "appropriate" are investigated by Meyer and Schreiber [1969] and Hartman [1972a]. Related concepts using L^2 norms instead of supremum norms appear in Kahane [1962; Chap. VI] and [KS; Chap. XII].

The study initiated by Helson and Kahane is elaborated on by Méla [1969] who ties up the theory with the theory of Sidon sets in $\mathbb{R}$. Many of Méla's results are forerunners of Déchamps-Gondim's results proved in Chapter 8 and also of her work on topological Sidon sets in $\mathbb{R}$ which appears in Déchamps-Gondim [1970b] and [1972]. See also her paper [1973]. These results are treated in Meyer's book [M; Chap. VI]. The general treatment there is similar to our Chapter 8, but the details are somewhat different.

Sidon sets in $\mathbb{R}^n$ are studied by Varopoulos [1970].

10.12 CONVERGENCE OF LACUNARY SERIES. A history of convergence of Hadamard lacunary series is given in the Notes to Chapter V of [Z; Vol. I, pp. 378-380]. See also Gaposhkin [1967b]. Papers concerning the range of a lacunary series are Weiss [1959] and Kahane, Weiss and Weiss [1963].

10.13 LACUNARITY FOR COMPACT NON-ABELIAN GROUPS. Section 37 in [HR; Vol. II] contains an introduction to this topic. For a history up to about 1969, see the Notes to that section and also 37.23 - 37.24 [HR]. Ragozin [1972b; Cor. 7.3] has shown that if G is a compact connected semisimple Lie group, then its dual object Σ contains no infinite Sidon sets. Price [1971] has shown that if $G = SU(2)$, then Σ does not even contain infinite $\Lambda(p)$ sets for any $p > 1$. Rider [1974] proves the same result for $G = SU(n)$. He also proves that if $G = U(n)$ and if E is a $\Lambda(p)$ set in Σ for some $p > 1$, then the degrees of the representations of E must be bounded. Earlier Cecchini [1972] proved that if G is a compact Lie group and if E is a $\Lambda(4)$ set in Σ, then the degrees of the representations of E must be bounded. In particular, E must be finite if G is semisimple. Dunkl and Ramirez [1971b] use Cecchini's results to prove that the union of two Sidon sets in Σ is a Sidon set provided G is a compact Lie group. Bożejko [1974a] shows that the union of two Sidon sets in Σ is a Sidon set provided the degrees of their representations are bounded.

Sidon and $\Lambda(p)$ sets in Σ are characterized by Figà-Talamanca and Rider [1966] and by Dunkl and Ramirez [1972b]. Baldoni [1973] constructs non-Sidon sets that are $\Lambda(p)$ sets for all p, $1 < p < \infty$, in the dual object Σ of certain products of unitary groups. Sidon sets that are peak sets are shown to be strong peak sets by Dunkl and Ramirez [1971a]. Sidon sets also appear in the work of Akemann [1968; Theorem 4].

10.14 CENTRAL LACUNARITY. The theory of central lacunary sets began with the 1970 thesis of W. A. Parker; see Parker [1972]. In this theory, algebras of functions and measures are replaced by their centers. The consequences of this little change are startling. For example, Parker proved that central Sidon sets need not be central $\Lambda(4)$ sets. In fact, there exist central Sidon sets that are not central $\Lambda(p)$ for any $p > 0$; see Rider [1972b]. Rider also gives an example of two central Sidon sets whose union is not a central Sidon set. Dunkl and Ramirez [1971a] showed that Σ is not a central Sidon set unless G is finite. Ragozin [1972a] proved that all central Sidon sets must be finite if G is a compact connected simple Lie group. Rider [1972b] sharpened this observation to the following: If G is compact and connected, then all central Sidon sets in Σ are finite if and only if G is a semisimple Lie group. Compare with Ragozin [1972b; Theorem 7.2]. Ragozin [1974] extends his work to "zonal Sidon sets" for homogeneous spaces.

Central $\Lambda(p)$ sets are studied by Rider [1972a] and Cecchini [1972]. Rider shows that Σ has an infinite central $\Lambda(2)$ set if G is a compact infinite connected Lie group. He also studies central $\Lambda(p)$ sets for $G = U(n)$ and $SU(n)$.

Central Sidon sets of bounded representation type are studied by Dunkl and Ramirez [1971b]. Benke [1974a] investigates the structure of central $\Lambda(p)$ sets in terms of the hypergroup structure of Σ. In particular, he shows that for $p > 2$, central $\Lambda(p)$ sets cannot contain arbitrarily long "arithmetic progressions".

Local Sidon and local $\Lambda(p)$ sets in Σ are investigated by Picardello [1974], Price [1974a] and [1974b], and Rider [1974].

It should not be surprising that the study of central lacunarity is related to the study of central measures. Considerable progress in this area has been made by Ragozin [1972a], [1972b] and [1974], Ragozin and Rothschild [1972], Rider [1970], [1971] and [1973], Mosak and Moskowitz [1971] and [1974], and Dunkl [1973].

10.15 LACUNARITY FOR DISCRETE NONCOMMUTATIVE GROUPS. For an infinite discrete group G, let $A(G)$ and $B(G)$ be the Fourier and Fourier-Stieltjes algebras which were defined and studied by Eymard [1964] and which have since been extensively studied. Let $VN(G)$ be the von Neumann algebra generated by all left convolution operators on $\ell^2(G)$, i.e. the dual space of $A(G)$. Let $c_{00}(G)$ consist of all functions on G with finite support. Figà-Talamanca and Picardello [1973] define a subset E of G to be a Sidon set provided

$$\|f\|_1 \leqq \varkappa \|f\|_{VN} \quad \text{for all} \quad f \in c_{00}(G)$$

and some constant $\varkappa > 0$. They also define E to be a Leinert set provided

$$\|f\|_{VN} \leqq \varkappa \|f\|_2 \quad \text{for all} \quad f \in c_{00}(G)$$

and some constant $\varkappa > 0$. Such sets were studied by Leinert [1970] and [1974], who showed that infinite Leinert sets exist in any free group with two generators. Figà-Talamanca and Picardello [1973] used Leinert's result to show that if G contains a free subgroup on two generators, then there exist multipliers of $A(G)$ that do not belong to $B(G)$. An alternative proof is given by Michele and Soardi [1974]. Picardello [1973] defines $\Lambda(q)$ sets in this setting for $1 < q \leq \infty$ in such a way that the class of $\Lambda(\infty)$ sets is precisely the class of Leinert sets. He proves that Sidon sets are $\Lambda(q)$ sets for $1 < q < \infty$, and he proves that every infinite subset of G contains an infinite $\Lambda(4)$ set. On the other hand, Figà-Talamanca and Picardello [1973] noted that Leinert [$= \Lambda(\infty)$] sets never contain infinite Sidon subsets! Compare Corollary 2.9. Božejko [1973], [1974d] establishes the existence of infinite $\Lambda(q)$ sets in every infinite discrete group for $1 < q < \infty$. Whether infinite Sidon sets exist in such groups remains unresolved. Božejko [1974b] studies $\Lambda(q)$ sets further and sharpens Leinert's original results.

Božejko [1974c] considers a discrete FC group G and its dual object E as defined by Thoma. A space $L^1(E)$ is defined; its Gelfand space $G^\sim$ can be identified with the set of conjugacy classes of elements of G. Sidon sets and $\Lambda(q)$ sets are defined in $G^\sim$ so that these notions reduce to the usual ones when G is abelian. Božejko observes that these Sidon sets need not be $\Lambda(q)$ sets and he characterizes the Sidon sets that are $\Lambda(q)$ sets.

Cygan [1974] studies Sidon sets in noncommutative locally compact groups.

SOME OPEN QUESTIONS

1. Is every Sidon set a Stechkin set? See 9.2, 3.7, and Corollary 2.19.

2. Given a Sidon set E, what properties does its closure $\overline{E}$ in the Bohr compactification possess? See the discussion after Theorem 4.9.

3. Are Λ-sets necessarily Sidon sets? See 6.7.

4. Does the Malliavin-Brameret and Malliavin theorem hold if p is not prime? See 6.12(c).

5. Is the union of two UC-sets again a UC-set? See 6.15.

6. Is every uncountable Sidon set the finite union of sets tending to infinity? See 9.2.

7. Are $\Lambda(4)$ sets in the dual of $\mathbb{Z}(p)^{\aleph_0}$ necessarily $\Lambda(q)$ sets for some $q > 4$? See 10.3.

8. If E has the property that, given $\mu \in M(G)$ where $\hat{\mu} \in c_0(X)$, we have $\hat{\mu}|_E = \hat{f}|_E$ for some $f \in L^1(G)$, must E be a Sidon set? See 10.9.

9. Do infinite Sidon sets exist in every infinite discrete noncommutative group? See 10.15.

BOOKS

[DR] Dunkl, C. F. and D. E. Ramirez, Topics in Harmonic Analysis, Appleton-Century-Crofts, New York, 1971.

[E] Edwards, R. E., Fourier Series: A Modern Introduction, Holt, Rinehart and Winston, New York, 1967, two volumes.

[HR] Hewitt, E. and K. A. Ross, Abstract Harmonic Analysis, Springer-Verlag, Heidelberg, 1963, 1970, two volumes.

[K] Kahane, J.-P., Séries de Fourier absolument convergentes, Springer-Verlag, Heidelberg, 1970.

[Ka] Kahane, J.-P., Some Random Series of Functions, Heath and Co., Lexington, Mass., 1968.

[KS] Kahane, J.-P. and R. Salem, Ensembles parfaits et Séries trigonométriques, Actualités sci. ind. 1301, Hermann, Paris, 1963.

[Kz] Katznelson, Y., An Introduction to Harmonic Analysis, John Wiley & Sons, New York, 1968.

[LP] Lindahl, L.-Å. and F. Poulsen, Thin Sets in Harmonic Analysis, Marcel Dekker, Inc., New York, 1971.

[M] Meyer, Y., Algebraic Numbers and Harmonic Analysis, North-Holland Publishing Co., 1972.

[R] Rudin, W., Fourier Analysis on Groups, Interscience Publishers (John Wiley & Sons), New York, 1962.

[Z] Zygmund, A., Trigonometric Series, Cambridge Press, 1959, two volumes.

OTHER REFERENCES

Akemann, C. A.

[1968] Invariant subspaces of $C(G)$, Pacific J. Math. 27 (1968), 421-424. MR38-5996

Bachelis, G. F.

[1971] On the ideal of unconditionally convergent Fourier series in $L_p(G)$, Proc. Amer. Math. Soc. 27 (1971), 309-312. MR42-6523

Bachelis, G. F. and S. E. Ebenstein

[1974] On $\Lambda(p)$ sets, Pacific J. Math. 53 (1974). To appear.

Badé, W. G. and P. C. Curtis, Jr.

[1966] Embedding theorems for commutative Banach algebras, Pacific J. Math. 18 (1966), 391-409. MR34-1878

Baker, R. C.

[1970] A diophantine problem on groups, II, Proc. London Math. Soc. 21 (1970), 757-768. MR46-613

[1972] On Wiener's theorem on Fourier-Stieltjes coefficients and the gaussian law, Proc. London Math. Soc. 25 (1972), 525-542. MR46-5929

Baldoni, M. W.

[1973] An example of a lacunary set for noncommutative group, Notices Amer. Math. Soc. 20 (1973), A433.

Benke, G.

[1971] Sidon sets and the growth of L^p norms, University of Maryland thesis, 1971.

[1972] Arithmetic structure and lacunary Fourier series, Proc. Amer. Math. Soc. 34 (1972), 128-132. MR46-614

[1974a] On the hypergroup structure of central $\Lambda(p)$ sets, Pacific J. Math. 50 (1974), 19-27.

[1974b] An example in the theory of $\Lambda(p)$-sets. To appear.

[1974c] The absolute convergence of certain lacunary Fourier series, Michigan Math. J. 21 (1974), 107-113.

[1974d] On the absolute convergence of lacunary Fourier series of Lipschitz functions, Indiana Univ. Math. J. 24 (1974/75), 265-269.

Blei, R. C.

[1972a] A note on some characterizations of Sidon sets, Proc. Amer. Math. Soc. 35 (1972), 303-304. MR46-2357

[1972b] On trigonometric series associated with separable, translation invariant subspaces of $L^\infty(G)$, Trans. Amer. Math. Soc. 173 (1972), 491-499. MR47-2269

[1973] On subsets with associated compacta in discrete abelian groups, Proc. Amer. Math. Soc. 37 (1973), 453-455. MR47-2274

[1974a] A simple diophantine condition in harmonic analysis, Studia Math. 52. To appear.

[1974b] A tensor approach to interpolation phenomena in discrete abelian groups, Proc. Amer. Math. Soc. To appear.

[1974c] On Fourier Stieltjes transforms of discrete measures, Math. Scand. To appear.

[1974d] Some thin sets in discrete abelian groups, Trans. Amer. Math. Soc. 193 (1974), 55-65.

Bonami, A.

[1968] Construction d'opérateurs de convolution sur le groupe D^{∞}, C. R. Acad. Sci. Paris 266A (1968), 864-866. MR37-6686

[1968a] Ensembles $\Lambda(p)$ dans le dual de D^{∞}, Ann. Inst. Fourier (Grenoble) 18, fasc. 2 (1968), 193-204. MR40-3181

[1970] Étude des coefficients de Fourier des fonctions de $L^p(G)$, Ann. Inst. Fourier (Grenoble) 20, fasc. 2 (1970), 335-402. MR44-727

Bonami, A. and Y. Meyer

[1969] Propriétés de convergence de certaines séries trigonométriques, C. R. Acad. Sci. Paris 269A (1969), 68-70. MR39-4584

Bożejko, M.

[1973] The existence of $\Lambda(p)$ sets in discrete noncommutative groups, Boll. Unione Mat. Ital. 8 (1973), 579-582.

[1974a] Sidon sets in dual object of compact group, Colloq. Math. 30 (1974), 137-141.

[1974b] On $\Lambda(p)$ sets with minimal constant in discrete noncommutative groups, Proc. Amer. Math. Soc. To appear.

[1974c] Sidon sets in discrete FC groups, Bull. Acad. Polon. Sci. To appear.

[1974d] A remark to my paper "The existence of $\Lambda(p)$ sets in discrete noncommutative groups", Boll. Unione Mat. Ital. To appear.

Bożejko, M. and T. Pytlik

[1972] Some types of lacunary Fourier series, Colloq. Math. 25 (1972), 117-124 and 164. MR46-4106

Brown, G.

[1974] Riesz products and generalized characters, Proc. London Math. Soc. To appear.

Cecchini, C.

[1972] Lacunary Fourier series on compact Lie groups, J. Functional Anal. 11 (1972), 191-203.

Chaney, R. W.

[1969] On uniformly approximable Sidon sets, Proc. Amer. Math. Soc. 21 (1969), 245-249. MR39-715

[1971] Note on Fourier-Stieltjes transforms of continuous and absolutely continuous measures, Rocz. Pols. Tow. Mat.: Prace Mat. 15 (1971), 147-149.

Cygan, J.

[1974] Riesz products on noncommutative groups, Studia Math. 51 (1974), 115-123.

Déchamps-Gondim, M.

[1970a] Compacts associés à un ensemble de Sidon, C. R. Acad. Sci. Paris 271A (1970), 590-592. MR42-6526

[1970b] Sur les ensembles de Sidon topologiques, C. R. Acad. Sci. Paris 271A (1970), 1247-1249. MR43-823

[1972] Ensembles de Sidon topologiques, Ann. Inst. Fourier (Grenoble) 22, fasc. 3 (1972), 51-79.

[1973] Interpolation lineaire approchée des fonctions bornées définies sur un ensemble de Sidon, Colloq. Math. 28 (1973), 255-259.

Dressler, R. E., W. Parker and L. Pigno

[1973] Sidon sets and small p sets, Quart. J. Math. Oxford (2) 24 (1973), 79-80. MR47-7336

Dressler, R. E. and L. Pigno

[1974a] On strong Riesz sets, Colloq. Math. 29 (1974), 157-158.

[1974b] Rosenthal sets and Riesz sets, Duke Math. J. 41 (1974), 675-677.

[1974c] Sets of uniform convergence and strong Riesz sets, Math. Annalen 211 (1974), 227-231.

[1975] A remark on "Rosenthal sets and Riesz sets", Kansas State Univ. Technical Report 51, 1975.

Drury, S. W.

[1970] Sur les ensembles de Sidon, C. R. Acad. Sci. Paris 271A (1970), 162-163. MR42-6530

[1972] Unions of sets of interpolation, Springer Lecture Notes 266 (1972), (1971 Maryland Conference), 23-33.

[1974] The Fatou-Zygmund property for Sidon sets, Bull. Amer. Math. Soc. 80 (1974), 535-538.

[1974a] Birelations and Sidon sets. To appear.

Dunkl, C. F.

[1973] Structure hypergroups for measure algebras, Pacific J. Math. 47 (1973), 413-425.

Dunkl, C. F. and D. E. Ramirez

[1971a] Sidon sets on compact groups, Monatsh. für Math. 75 (1971), 111-117. MR46-5938

[1971b] Central Sidon sets of bounded representation type, Notices Amer. Math. Soc. 18 (1971), A1101.

[1972a] C^*-algebras generated by Fourier-Stieltjes transforms, Trans. Amer. Math. Soc. 164 (1972), 435-441. MR46-9646

[1972b] Characterizations of Sidon sets and Λ_p sets on compact groups, Notices Amer. Math. Soc. 19 (1972), A436-437.

[1973] Sections induced from weakly sequentially complete spaces, Studia Math. 49 (1973), 95-97.

Ebenstein, S. E.

[1972] $\Lambda(p)$ sets and Sidon sets, Proc. Amer. Math. Soc. 36 (1972), 619-620. MR46-9649

[1974] $\Lambda(p)$ sets and the exact majorant property, Proc. Amer. Math. Soc. 42 (1974), 533-534.

Eberlein, W.

[1955] The point spectrum of weakly almost periodic functions, Mich. Math. J. 3 (1955-56), 137-139. MR18-583

Edwards, R. E.

[1973] Lipschitz conditions and lacunarity, J. Austral. Math. Soc. 16 (1973), 272-277.

Edwards, R. E., E. Hewitt and K. A. Ross

[1972a] Lacunarity for compact groups, I, Indiana J. Math. 21 (1972), 787-806. MR45-6981

[1972b] Lacunarity for compact groups, II, Pacific J. Math. 41 (1972), 99-109. MR47-3911

[1972c] Lacunarity for compact groups, III, Studia Math. 44 (1972), 429-476.

Edwards, R. E. and K. A. Ross

[1973] Helgason's number and lacunarity constants, Bull. Austral. Math. Soc. 9 (1973), 187-218.

[1974] p-Sidon sets, J. Functional Anal. 15 (1974), 404-427.

Eymard, P.

[1964] L'algèbre de Fourier d'un groupe localement compact, Bull. Soc. math. France 92 (1964), 181-236. MR37-4208

Figà-Talamanca, A.

[1964] Multipliers of p-integrable functions, Bull. Amer. Math. Soc. 70 (1964), 666-669. MR29-2327

[1970] An example in the theory of lacunary Fourier series, Boll. Unione Mat. Ital. 3 (1970), 375-378. MR42-762

Figa-Talamanca, A. and M. Picardello

[1973] Multiplicateurs de A(G) qui ne sont pas dans B(G), C. R. Acad. Sci. Paris 277A (1973), 117-119.

Figa-Talamanca, A. and D. Rider

[1965] Translation invariant operators in L^p, Duke Math. J. 32 (1965), 495-501. MR31-6095

[1966] A theorem of Littlewood and lacunary series for compact groups, Pacific J. Math. 16 (1966), 505-514. MR34-6444

Fournier, J.

[1969] Extensions of a Fourier multiplier theorem of Paley, Pacific J. Math. 30 (1969), 415-431. MR41-2301

[1973a] Fourier coefficients after gaps, J. Math. Anal. Appl. 42 (1973), 255-270. MR48-795

[1973b] Local complements to the Hausdorff-Young theorem, Mich. Math. J. 20 (1973), 263-276. MR47-7333

[1974] An interpolation problem for coefficients of H^∞ functions, Proc. Amer. Math. Soc. 42 (1974), 402-408.

Gaposhkin, V. F.

[1967a] Asymptotically best lacunary systems, Izv. Akad. Nauk SSSR 31 (1967), 1011-1026 (Russian); Math. USSR Izv. 1 (1967), 967-982 (English). MR38-3684

[1967b] A question on the absolute convergence of lacunary series, Izv. Akad. Nauk SSSR 31 (1967), 1271-1288 (Russian); Math. USSR Izv. 1 (1967), 1217-1234 (English). MR36-5598

Glicksberg, I.

[1974] Interpolation by cones, Studia Math. 49 (1974), 235-251.

Glicksberg, I. and I. Wik

[1972] The range of Fourier-Stieltjes transforms of parts of measures, Springer Lecture Notes 266 (1972), (1971 Maryland Conference), 73-77.

Graham, C. C.

[1973a] Two remarks on the Fourier-Stieltjes transforms of continuous measures, Colloq. Math. 27 (1973), 297-299.

[1973b] Sur un théorème de Katznelson et McGehee, C. R. Acad. Sci. Paris 276A (1973), 37-40. MR47-3910

[1974] A Riesz product proof of the Wiener-Pitt theorem, Proc. Amer. Math. Soc. 44 (1974), 312-314.

Gregory, M. B.

[1972] p-Helson sets, $1 < p < 2$, Israel J. Math. 12 (1972), 356-368. MR47-7338

Hahn, L.-S.

[1972] Fourier series with gaps. Unpublished.

Hartman, S.

[1961] On interpolation by almost periodic functions, Colloq. Math. 8 (1961), 99-101. MR23-A3416

[1968] Interpolation und Gleichverteilung in Bohr's Kompaktifizierung, Colloq. Math. 19 (1968), 111-115. MR38-2529

[1972a] The method of Grothendieck-Ramirez and weak topologies in C(T), Studia Math. 44 (1972), 181-197. MR47-7334

[1972b] Interpolation par les mesures diffuses, Colloq. Math. 26 (1972), 339-343.

[1974] Remark on interpolation by L-almost periodic functions, Colloq. Math. 30 (1974), 133-136.

Hartman, S., J.-P. Kahane and C. Ryll-Nardzewski

[1965] Sur les ensembles d'interpolation, Bull. Acad. Polon. Sci. 13 (1965), 625-626. MR33-1645

Hartman, S., S. Rolewicz and C. Ryll-Nardzewski

[1967] Ultra-Kroneckerian sets, Colloq. Math. 16 (1967), 225-229. MR35-5859

Hartman, S. and C. Ryll-Nardzewski

[1963] Almost periodic extensions of functions, Bull. Acad. Polon. Sci. 11 (1963), 427-429. MR29-6253

[1964] Almost periodic extensions of functions, Colloq. Math. 12 (1964), 23-39. MR29-5057

[1966] Almost periodic extensions of functions, II, Colloq. Math. 15 (1966), 79-86. MR33-7794

[1967] Almost periodic extensions of functions, III, Colloq. Math. 16 (1967), 223-224. MR35-4676

[1971] Quelques résultats et problèmes en algèbre des mesures continues, Colloq. Math. 22 (1971), 271-277. MR44-1993

Helson, H. and J.-P. Kahane

[1965] A Fourier method in Diophantine problems, J. Analyse Math. 15 (1965), 245-262. MR31-5856

Herz, C.

[1972] Drury's lemma and Helson sets, Studia Math. 42 (1972), 205-219. MR46-5939

Hewitt, E.

[1968] A generalization of certain polynomials of Frédéric Riesz, 287-296 in Orthogonal expansions and their continuous analogues, Southern Illinois University Press, 1968. MR39-1906

Hewitt, E. and H. S. Zuckerman

[1959] Some theorems on lacunary Fourier series, with extensions to compact groups, Trans. Amer. Math. Soc. 93 (1959), 1-19. MR21-7400

[1966] Singular measures with absolutely continuous convolution squares, Proc. Camb. Philos. Soc. 62 (1966), 399-420. Corrigendum, same Proc. 63 (1967), 367-368. MR33-1655, MR34-8097

Horn, A.

[1955] A characterization of unions of linearly independent sets, J. London Math. Soc. 30 (1955), 494-496. MR17-135

Johnson, G. W.

[1974] Theorems on lacunary sets, especially p-Sidon sets. To appear.

Johnson, G. W. and G. S. Woodward

[1974] On p-Sidon sets, Indiana Univ. Math. J. 24 (1974/75), 161-167.

Kahane, J.-P.

[1957] Sur les fonctions moyennes-périodiques bornées, Ann. Inst. Fourier (Grenoble) 7 (1957), 293-314. MR21-1489

[1962] Pseudo-périodicité et séries de Fourier lacunaires, Ann. sci. Ecole Norm. Sup. 79 (1962), 93-150. MR27-4019

[1964] Lacunary Taylor and Fourier series, Bull. Amer. Math. Soc. 70 (1964), 199-213. MR29-223

[1966] Ensembles de Ryll-Nardzewski et ensembles de Helson, Colloq. Math. 15 (1966), 87-92. MR32-8041

Kahane, J.-P., M. Weiss and G. Weiss

[1963] On lacunary power series, Arkiv fȍr Mat. 5 (1963), 1-26. MR32-7716

Kaufman, R.

[1969] A random method for lacunary series, J. Analyse Math. 22 (1969), 171-175. MR40-623

[1970] A remark on Sidon sets, Rev. Un. Mat. Argentina 25 (1970), 105-107.

[1971] Remark on Fourier-Stieltjes transforms of continuous measures, Colloq. Math. 22 (1971), 279-280. MR44-1994

Keogh, F.

[1965] Riesz products, Proc. London Math. Soc. (3) 14a (1965), 174-182. MR32-4456

Leinert, M.

[1970] Convoluteurs de groupes discrets, C. R. Acad. Sci. Paris 271A (1970), 630-631. MR42-6522

[1974] Faltungsoperatoren auf gewissen diskreten Gruppen, Studia Math. 52 (1974), 149-158.

López, J.

[1974] Fatou-Zygmund properties on groups, University of Oregon thesis, 1975.

Malliavin-Brameret, M.-P. and P. Malliavin

[1967] Caractérisation arithmétique d'une classe d'ensembles de Helson, C. R. Acad. Sci. Paris 264A (1967), 192-193. MR35-669

Méla, J.-F.

[1964] Suites lacunaires de Sidon, ensembles propres et points exceptionnels, Ann. Inst. Fourier (Grenoble) 14, fasc. 2 (1964), 533-538. MR31-2566

[1968a] Sur certains ensembles exceptionnels en analyse de Fourier, Ann. Inst. Fourier (Grenoble) 18 (1968), 32-71.

[1968b] Sur les ensembles d'interpolation de C. Ryll-Nardzewski et de S. Hartman, Studia Math. 29 (1968), 167-193. MR36-5616

[1969] Approximation Diophantienne et ensembles lacunaires, Bull. Soc. math. France, Memoire 19 (1969), 26-54.

Meyer, Y.

[1968a] Endomorphismes des idéaux fermés de $L^1(G)$, classes de Hardy et séries de Fourier lacunaires, Ann. sci. École Norm. Sup. 1 (1968), 499-580. MR39-1910

[1968b] Spectres des mesures et mesures absolument continues, Studia Math. 30 (1968), 87-99. MR37-3281

[1972] Recent advances in spectral synthesis, Springer Lecture Notes 266 (1972), (1971 Maryland Conference), 239-253.

Meyer, Y. and J.-P. Schreiber

[1969] Quelques fonctions moyennes-périodiques non bornées, Ann. Inst. Fourier (Grenoble) 19, fasc. 1 (1969), 231-236. MR40-3176

Michele, L. de and P. M. Soardi

[1974] A noncommutative extension of Helson's translation lemma, Boll. Unione Mat. Ital. 9 (1974), 800-806.

Mosak, R. D. and M. Moskowitz

[1971] Central idempotents in measure algebras, Math. Z. 122 (1971), 217-222. MR45-9062

[1974] Central idempotents in group algebras, Proc. Amer. Math. Soc. To appear.

Orlov, E. V.

[1973] A class of lacunary trigonometric series, Mat. Zametki 14 (1973), 781-788 (Russian); Math. Notes 14 (1973), 1006-1010 (English).

Padé, O.

[1973] Sur le spectre d'une classe de produits de Riesz, C. R. Acad. Sci. Paris 276A (1973), 1453-1455. MR47-7306

Parker, W. A.

[1972] Central Sidon and central Λ_p sets, J. Austral. Math. Soc. 14 (1972), 62-74. MR47-9178

Pedemonte, L.

[1974] Sets of uniform convergence, Colloq. Math. To appear.

Peyrière, J.

[1973] Sur les produits de Riesz, C. R. Acad. Sci. Paris 276A (1973), 1417-1419. MR47-5512

Picardello, M. A.

[1973] Lacunary sets in discrete noncommutative groups, Boll. Unione Mat. Ital. 8 (1973), 494-508.

[1974] Random Fourier series on compact non commutative groups, Canad. J. Math. To appear.

Pigno, L.

[1974a] Approximations to the norm of the singular part of a measure, Kansas State Univ. Technical Report 44, 1974.

[1974b] A variant of the F. and M. Riesz theorem, J. London Math. Soc. To appear.

Pigno, L. and S. Saeki

[1973] Measures whose transforms vanish at infinity, Bull. Amer. Math. Soc. 79 (1973), 800-801. MR47-5522

[1974] Almost periodic functions and certain lacunary sets, Kansas State Univ. Technical Report 42, 1974. See also "On the spectra of almost periodic functions", Indiana Univ. Math. J. To appear.

Price, J. F.

[1970] Some strict inclusions between spaces of L^p-multipliers, Trans. Amer. Math. Soc. 152 (1970), 321-330. MR43-7923

[1971] Non ci sono insiemi infiniti di tipo $\Lambda(p)$ per SU_2, Boll. Unione Mat. Ital. 4 (1971), 879-881. MR45-8772

[1974a] On local central lacunary sets for compact Lie groups, Monatsh. für Math. To appear.

[1974b] Local Sidon sets and uniform convergence of Fourier series, Israel J. Math. 17 (1974), 169-175.

Rado, R.

[1962] A combinatorial theorem on vector spaces, J. London Math. Soc. 37 (1962), 351-353. MR26-3708

Ragozin, D. L.

[1972a] Central measures on compact simple Lie groups, J. Functional Anal. 10 (1972), 212-229.

[1972b] Vector measure algebras and central measures on semisimple Lie groups. Unpublished.

[1974] Zonal measure algebras on isotropy irreducible homogeneous spaces, J. Functional Anal. To appear.

Ragozin, D. L. and L. P. Rothschild

[1972] Central measures on semisimple Lie groups have essentially compact support, Proc. Amer. Math. Soc. 32 (1972), 585-589. MR45-466

Ramirez, D. E.

[1968] Uniform approximation by Fourier-Stieltjes transforms, Proc. Camb. Philos. Soc. 64 (1968), 615-623. MR37-696

[1973] Remark on Fourier-Stieltjes transforms of continuous measures, Colloq. Math. 27 (1973), 81-82.

Rider, D.

[1966a] Gap series on groups and spheres, Canad. J. Math. 18 (1966), 389-398. MR32-8047

[1966b] A relation between a theorem of Bohr and Sidon sets, Bull. Amer. Math. Soc. 72 (1966), 558-561. MR33-508

[1970] Central idempotent measures on unitary groups, Canad. J. Math. 22 (1970), 719-725. MR41-8929

[1971] Central idempotent measures on SIN groups, Duke Math. J. 38 (1971), 187-191. MR42-6518

[1972a] Norms of characters and central Λ_p sets for U(n), Springer Lecture Notes 266 (1972), (1971 Maryland Conference), 287-294.

[1972b] Central lacunary sets, Monatsh. für Math. 76 (1972), 328-338.

[1973] Central idempotent measures on compact groups, Trans. Amer. Math. Soc. 186 (1973), 459-479.

[1974] SU(n) has no infinite local Λ_p sets. To appear.

Rosenthal, H. P.

[1967] On trigonometric series associated with weak* closed subspaces of continuous functions, J. Math. Mech. 17 (1967), 485-490. MR35-7064

Ross, K. A.

[1972] Sur les compacts associés à un ensemble de Sidon, C. R. Acad. Sci. Paris 275A (1972), 183-185. MR46-5940

[1973] Fatou-Zygmund sets, Proc. Camb. Philos. Soc. 73 (1973), 57-65. MR46-9651

Rudin, W.

[1960] Trigonometric series with gaps, J. Math. Mech. 9 (1960), 203-227. MR22-6972

Ryll-Nardzewski, C.

[1963] Remarks on interpolation by periodic functions, Bull. Acad. Polon. Sci. 11 (1963), 363-366. MR27-5078

[1964] Concerning almost periodic extensions of functions, Colloq. Math. 12 (1964), 235-237. MR30-3344

Stafney, J. D.

[1969] Approximation of W_p-continuity sets by p-Sidon sets, Mich. Math. J. 16 (1969), 161-176. MR40-632

Stechkin, S. B.

[1956] On absolute convergence of Fourier series, Izv. Akad. Nauk SSSR 20 (1956), 385-412 (Russian). MR18-126

Stromberg, K.

[1968] Large families of singular measures having absolutely continuous convolution squares, Proc. Camb. Philos. Soc. 64 (1968), 1015-1022. MR37-6693

Strzelecki, E.

[1963] On a problem of interpolation by periodic and almost periodic functions, Colloq. Math. 11 (1963), 91-99. MR28-3294

Suslikova, O. G.

[1973] On trigonometric Fourier series with lacuna, Vestnik Moskov. Univ. Mat. 1973, No. 6, 46-55 (Russian); Moscow Univ. Math. Bull. 28 (1973), 92-99 (English).

Varopoulos, N. Th.

[1967] Sidon sets in R^n, Math. Scand. 27 (1970), 39-49. MR42-8184

[1970] Tensor algebras and harmonic analysis, Acta Math. 119 (1967), 51-112. MR39-1911

Weiss, M.

[1959] Concerning a theorem of Paley on lacunary power series, Acta Math. 102 (1959), 225-238. MR22-8109

Wells, B. B., Jr.

[1971] Sets of interpolation for multipliers, Bull. Amer. Math. Soc. 77 (1971), 220-222. MR42-4970

[1973] Restrictions of Fourier transforms of continuous measures, Proc. Amer. Math. Soc. 38 (1973), 92-94. MR47-3913

Woodward, G. S.

[1974] p-Sidon sets and a uniform condition. To appear.

Zygmund, A.

[1930] On the convergence of lacunary trigonometric series, Fund. Math. 16 (1930), 90-107.

[1931] Quelques théorèmes sur les séries trigonométriques et celles de puissances, Studia Math. 3 (1931), 77-91.

[1932] On lacunary trigonometric series, Trans. Amer. Math. Soc. 34 (1932), 435-446.

[1948] On a theorem of Hadamard, Ann. Soc. Polon. Math. 21 (1948), 52-69; Errata, 357-358. MR10-186, MR11-20

SYMBOLS

INDEX